1969

This book may be kept

INTERNATIONAL SERIES OF MONOGRAPHS ON
PURE AND APPLIED BIOLOGY

Division: **ZOOLOGY**

GENERAL EDITOR: G. KERKUT

VOLUME 14

THE CONTROL
OF
CHROMATOPHORES

OTHER TITLES IN THE ZOOLOGY DIVISION

General Editor: G. A. KERKUT

OTHER DIVISIONS IN THE SERIES ON
PURE AND APPLIED BIOLOGY

BIOCHEMISTRY

BOTANY

MODERN TRENDS
IN PHYSIOLOGICAL SCIENCES

PLANT PHYSIOLOGY

THE CONTROL
OF
CHROMATOPHORES

M. FINGERMAN

DEPARTMENT OF ZOOLOGY
NEWCOMB COLLEGE
TULANE UNIVERSITY
NEW ORLEANS

A Pergamon Press Book

THE MACMILLAN COMPANY
NEW YORK
1963

This book is distributed by

THE MACMILLAN COMPANY · NEW YORK

pursuant to a special arrangement with

PERGAMON PRESS INC.

NEW YORK, N. Y.

Library of Congress Card Number 62–11547

Printed in Poland to the order of PWN—Polish Scientific Publishers
by the Scientific-Technical Printing House (DRP), Warsaw

To My Wife

TABLE OF CONTENTS

ACKNOWLEDGEMENTS

The author wishes to express his appreciation to the National Institute of Health for having supported his research program so generously.

The courtesy of the following publishers in permitting reproduction of copyright material is also acknowledged with gratitude: Academic Press, Almqvist and Wiksell, American Association for the Advancement of Science, American Institute of Biological Sciences, American Microscopical Society, Cambridge University Press, Commonwealth Scientific and Industrial Research Organization (Australia), Gustav Fischer Verlag, Rockefeller Institute, Royal Society of London, Society of Experimental Biology and Medicine, Springer Verlag, The Marine Biological Laboratory, The University of Chicago Press, University of Notre Dame Press; Wistar Institute, Zoological Society of Japan.

CHAPTER 1

CHROMATOPHORES AND COLOR CHANGES

COLOR changes in animals have commanded the attention of investigators since Aristotle, but interest in this subject has never been higher than it is currently. In 1948 Parker published a comprehensive volume on the control of chromatophores. His treatise stands as a landmark in this area of research and is evidence of the magnificent contributions he made. The monograph deals in the main with publications that appeared between 1910 and 1943; in the bibliography over 1200 references are cited.

Two significant reviews concerning color changes that were published in this century are those of Van Rynberk (1906) and Fuchs (1914). Since publication of Parker's volume, several reviews of the literature concerning color changes have appeared, but in the main these have emphasized restricted aspects of the subject (Waring and Landgrebe, 1950; Brown, 1952a; Scharrer, 1952a; Knowles and Carlisle, 1956; Pickford and Atz, 1957; Gersch, 1958; Carlisle and Knowles, 1959; Fingerman, 1959a; Scheer, 1960; Prosser and Brown, 1961).

Since the publication of Parker's volume many advances have been made in our understanding of the control of color changes, particularly among the invertebrates. This review will deal principally with work reported after the appearance of Parker's treatise. Earlier publications will be mentioned when necessary for full comprehension of current efforts.

TYPES OF CHROMATOPHORES

Cells containing pigment that can disperse or concentrate, thereby changing the tint of the organism in which they lie, are known

1

as chromatophores. Such effectors may be close to the surface of an organism or deep, surrounding various organs. The term color change is somewhat of a misnomer because in most cases the color of the animal does not change but merely its shade or tint. Furthermore, to be technically correct, we should not speak of black or white chromatophores being involved in color changes since white is all colors together and black is the absence of color. However, the idea that these pigments are a part of color change is so engrained in the literature that it would be a fruitless effort to attempt to rectify the situation at this time.

Well-developed, functional chromatophore systems are common among cephalopods, crustaceans, and poikilotherm vertebrates. Sporadic instances are found among other groups such as annelids, echinoderms, and insects. Two major types of chromatophores exist. One type is found in cephalopod molluscs and is a multicellular, complex organ. Other animal groups have unicellular chromatophores or groups of unicellular chromatophores so closely arranged that they appear to form a single color unit. Ballowitz (1914) termed such color units chromatosomes.

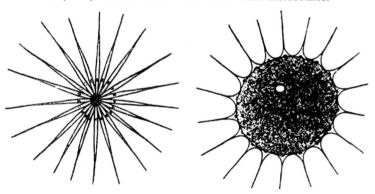

FIG. 1. Chromatophores of the squid *Loligo*. Left, retracted pigment; right, expanded pigment. (From Bozler, 1928.)

The chromatophores of cephalopods consist of a nucleus and a central pigment-containing sac surrounded by 6–20 or more radiating uninucleate smooth muscle fibers (Fig. 1). When the fibers are relaxed the pigment sac contributes essentially nothing to the coloration of the animal. When the muscle fibers contract, the pigment sac is pulled out to form a thin disc whose diameter may be from 2–60 times the relaxed diameter. The pigment sac appears to be elastic so that it returns to its original size when the muscle fibers relax (Bozler, 1928).

The generally accepted view concerning unicellular chromatophores is that they have a fixed cell outline and that the pigment migrates into and out of the processes. The earlier view was that unicellular chromatophores were amoeboid sending out pseudopodia when pigment dispersion occurred. Matthews (1931) was one of the first to help correct this notion. He detected the branched outline of chromatophores in which the pigment was maxi-

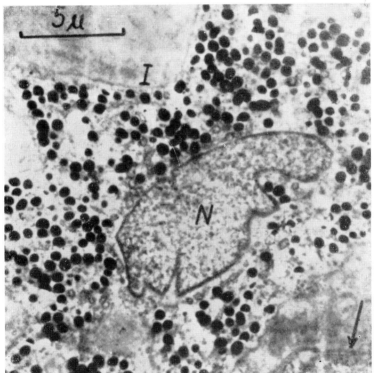

FIG. 2. A portion of a melanophore from *Lebistes* with concentrated melanin. I, inner cytoplasmic membrane; N, nucleus. Part of cell membrane at arrow. (From Falk and Rhodin, 1957.)

mally concentrated. Support for this contention can also be mustered from observations that the shape of a chromatophore is unchanged as its contained pigment migrates from the fully dispersed condition to the maximally concentrated state and back again to full dispersion (Spaeth, 1913; Perkins, 1928; Brown, 1935a). More recently, Falk and Rhodin (1957) studied pigment migration in the melanophores of the teleost *Lebistes reticulatus* by means of the electron microscope (Figs. 2, 3). Each melanophore is surrounded by a thick cell membrane, about 1200 Å. An inner cytoplasmic membrane, 80 Å in diameter, forms a sac that contains the nucleus, mitochondria, and pigment granules. Between the two cell membranes is a zone that is traversed by fibrils about 80 Å thick. These fibrils do not appear to be attached to either of the two cell membranes. The width of the fibrillar zone varies with the amount of pigment dispersion. In melanophores with maximally dispersed pigment the width is about 0.5 μ, and in cells with maximally concentrated pigment 3.0 μ. The fibrils are thought to be contractile structures which form a meshwork around the inner sac. When the fibrils contract the size of the sac would decrease and the pigment mass would become concentrated. Dispersion would result from relaxation of the fibrils.

Chromatophores are classified according to the pigment or pigments contained therein. If the pigment is brown or black, the cells are referred to as melanophores. Yellow chromatophores are called xanthophores; red ones, erythrophores. Cells that contain guanine or guanine-like substances are called guanophores. When the guanine consists of fine granules that can migrate the term leucophore is usually employed. In some cells, called iridophores or iridocytes, the white pigment consists of plate-like crystals that give the bearer an irridescent sheen. Iridophores are technically not chromatophores according to the definition supplied above, because the pigment in these cells does not migrate. The term lipophore has been used as a general term for chromatophores in which the pigment is soluble in organic solvents. Yellow and red pigments typically fall into this category.

In crustaceans and most vertebrates the chromatophores are

highly branched cells. However, in insects epidermal cells may also function as chromatophores (Giersberg, 1928; Key and Day, 1954a, b). In *Carausius morosus*, the walking stick, one finds in the distal portion of the epidermal cells a homogeneously distributed yellow-green pigment layer that is apparently immobile. When the insect darkens a dark brown-black pigment within

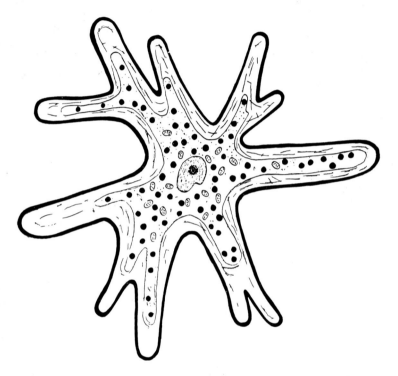

FIG. 3. Diagrammatic representation of the structure of a melanophore according to Falk and Rhodin (1957.)

these cells migrates from small concentrated masses in the proximal portion of the cell into the yellow-green pigment and disperses. At the same time an orange-red pigment found in small spherical masses in the yellow-green pigment layer disperses to form a continuous layer of pigment just proximal to the brown-black layer.

Key and Day (1954a, b) described an interesting type of chromatophore found in the grasshopper *Kosciuscola tristis* (Fig. 4). Each epidermal cell contains blue and brown granules. When the animal is dark the nucleus lies in the center of the cell, the brown

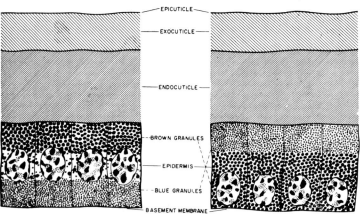

FIG. 4. Diagrammatic representation of hypodermal cells of *Kosciuscola tristis* showing the distribution of pigments in the dark phase (left) and in the blue phase (right). (From Key and Day, 1954a.)

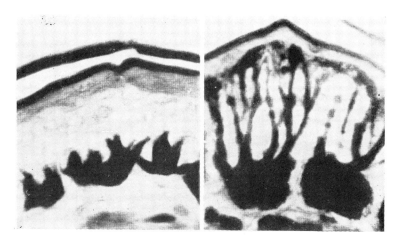

FIG. 5. Sections through scales of *Anolis carolinensis* showing the condition of the melanophores in the green phase (left) and brown phase (right). (From Kleinholz, 1938a.)

granules are dispersed in the distal half of the cell, and the blue granules are dispersed in the proximal half of the cell. When the grasshopper assumes its light phase, the nucleus migrates to the proximal portion of the cell where it lies close to the basement membrane and the positions of the blue and the brown pigments become interchanged. The displacement of the nucleus as the pigments migrate is intriguing.

Another interesting modification of the unicellular type of chromatophore is that found in the lizard *Anolis carolinensis* (Kleinholz, 1938a). The cell body of the chromatophore lies below an immobile light-colored pigment layer (Fig. 5). When the animal darkens, the pigment moves within processes of the melanophores to a position distal to the immobile pigment layer, thereby rendering the animal dark.

CHEMISTRY OF CHROMATOPHORAL PIGMENTS

The chemistry of chromatophoral pigments has been discussed in excellent fashion by Fox (1953) and Fox and Vevers (1960) for all animals and by Goodwin (1960) for crustaceans. The pigment in the melanophores of vertebrates and brachyuran decapods is melanin. This pigment is brown or black and can be extracted from tissues with boiling alkali. The term melanin applies to a variety of pigments and no exact definition can be given. Chemically the melanins are probably polymerized indole quinones formed by the action of tyrosinase on aromatic amino acids, especially tyrosine. The dark pigments in crustaceans aside from the crabs appear to be ommochromes rather than melanins.

The red and yellow chromatophoral pigments are carotenoids in all animals except cephalopods. These pigments are synthesized *de novo* only by plants. Animals must obtain carotenoids in their diet but can alter them by oxidation. Astaxanthin is probably the most common chromatophoral carotenoid in crustaceans (Brown, 1934). Astaxanthin also occurs in the erythrophores of a number of fishes (Goodwin, 1951). The pigment in the leucophores of *Rana pipiens* is guanine (Bagnara and Neidleman, 1958). The black pigment of the sea urchin *Diadema antillarum* gives reactions that are characteristic of melanin (Millott, 1950). Yellow, orange, and brownish-violet chromatophores occur in the cephalopod *Sepia officinalis*. The extracted pigments have the properties of ommochromes, including a reversible redox color change in solution from yellow-brown oxidized to wine red reduced. The different colors in the chromatophores could be due to different oxidation-reduction stages, to polymerization, or to conjugation with proteins of a single ommochrome (Schwinck, 1953, 1955). Cephalopod chromatophores apparently lack melanin. The skin pigments of *Octopus vulgaris* and *Eledone moschata* show the same reactions as the pigments of *Sepia*.

CLASSIFICATION OF CHROMATOPHORE RESPONSES

Color changes have been divided into two categories, morphological and physiological. The former implies a change in the quantity of pigment in an organism. Physiological color changes involve alteration upon appropriate stimulation in the degree of dispersion of the pigment granules in the chromatophores. Morphological color changes are usually evoked by maintaining an animal on a specific background for a number of weeks. Increase or decrease not only of the total quantity of pigment may be involved but also change in the number of chromatophores possessed by the animal. Physiological color changes may be evoked by a number of factors; the most important is light; temperature is second in importance. Factors of lesser importance are humidity, endogenous rhythms, psychic stimuli, and tactile stimuli.

Responses of chromatophores to light may be divided into two categories, primary and secondary. Primary color changes typically occur through routes other than the eyes, i.e. by direct action of light on the chromatophore, the more common type, or through an extraocular reflex, a response that involves either nervous or endocrine coordination between a stimulus received by a receptor other than the eye and the chromatophore. Primary responses, although typically associated with larval or embryonic pigment cells that become fully functional before the eyes are operative, may be exhibited by chromatophores of adults. Secondary responses depend upon the nature of the background and not on the intensity of light; the degree of pigment dispersion is determined by the ratio of the amount of light directly incident on the eye to the quantity impinging on the eye after reflection from the background. This ratio is known as "the albedo". In most adults the primary response is dominated by the secondary one.

Waring and Landgrebe (1950) revised the classification of responses of chromatophores to light as follows: (1) an uncoordinated nonvisual or dermal response which is independent of the eyes, central nervous system, and the pituitary in vertebrates, so that the chromatophores almost certainly behave as independent effectors; (2) a coordinated nonvisual response which is independent of the eyes but involves either nervous or pituitary coordination between a stimulus received by some receptor other than the pigment cells and the chromatophores themselves; (3) a secondary ocular response which results in melanin dispersion in specimens on a black background; and (4) a tertiary ocular response which results in melanin concentration when specimens are on a white background. The first two categories are merely subdivisions of the classical primary response, and the last two are subdivisions of the secondary response. In the opinion of the present author, division of the classical secondary response is unwarranted, and furthermore, the restricted definition of the term "secondary response" proposed by Waring and Landgrebe can lead only to confusion in the literature.

FUNCTIONAL SIGNIFICANCE OF COLOR CHANGES

The commonly ascribed functions of color changes are (1) protective coloration, (2) thermoregulation, and (3) displays associated with mating and parental behavior. Dispersion of dark pigments when an organism is put on a dark background obviously better adapts an organism for both aggression and survival. For example, Sumner (1935) demonstrated the value of protective coloration in the mosquito fish *Gambusia patruelis*. He placed pale and dark fish into black and into white tanks and found that predatory birds captured a smaller percentage of the light fish in a white container than dark ones and fewer dark fish in a black container than light ones.

Concomitant with the ability to change coloration for protection of the individual should be the ability to select the background that would assure maximum protection. Evidence in favor of this hypothesis was presented by Brown and Thompson (1937), who showed that in eight species of fresh-water fishes, those adapted to a black background tended to select black more frequently than fishes adapted to a white background. The rate of change of choice by the silver-mouthed minnow, *Ericymba buccata*, after a change of background was approximately the same as the rate of change in skin coloration. Brown (1939a) showed that the crayfish, *Orconectes immunis*, also has the ability to select between black and white backgrounds, always tending to select black over white.

Primary responses of chromatophores to light have been noticed in a wide variety of invertebrates and vertebrates. Quite often these responses are intimately associated with responses to temperature that have been interpreted as thermoregulatory. Brown and Sandeen (1948) determined the responses of the black and white chromatophores of the fiddler crab, *Uca pugilator*, to light and temperature. Both pigments dispersed with increased illumination. These pigments also showed a temperature response. The black pigment tended to concentrate as the temperature was raised above or lowered below about 15°C. White pigment, on the other hand, tended to disperse as the temperature increased above or decreased below about 20°C. The responses to high illumination and temperature may be thermoregulatory for protection of the protoplasm of the crabs. The optimal condition of the chromatophore system to produce dissipation of light and heat when the temperature is high would be, as is the actual case, dispersion of white pigment and concentration of black. Concentration of the black pigment tends to diminish the surface area that absorbs light, but dispersion of the white pigment tends to increase the area that reflects it. Pautsch (1953) found that the chromatophores of the zoea of the shrimp, *Crangon crangon*, exhibit only a primary response; the melanin dispersed with increased incident illumination.

The responses of the chromatophores of the grasshopper *Kosciuscola tristis* to temperature were described by Key and Day (1954a, b). Near 15°C the grasshopper is a dull black, and above 25°C a bright greenish-blue with intermediate tints between 15°–25°C. On clear days the insects become pale 2–3 hr after sunrise and then begin to turn dark again in the late afternoon. This color change is probably thermoregulatory, permitting the grasshopper to minimize the heating effects of the midday sun. Adult *Carausius morosus*, a phasmid, (Giersberg, 1928) and larval *Chaoborus plumicornis*, a dipteran, (Kopenec, 1949) also become pale with rise in temperature.

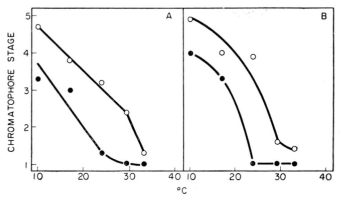

FIG. 6. Relationships between temperature and white pigment dispersion in *Palaemonetes pugio* (A) and *Palaemonetes paludosus* (B). Dots, black background; circles, white background. (From Fingerman and Tinkle, 1956.)

Brown, Sandeen, and Webb (1948) and Webb, Brown, and Graves (1952) studied the responses of the white pigment of *Palaemonetes vulgaris* to changes in intensity of illumination. In specimens on a black background, the white pigment dispersed with increased illumination. Fingerman and Tinkle (1956) observed the responses of the white chromatophores of two other species of this genus of prawn, *Palaemonetes pugio* and *Palaemonetes paludosus*, to light and temperature. The white pigment in both species dispersed with increase in total illumination as was observed by Brown, Sandeen, and Webb (1948) and Webb, Brown, and Graves (1952). However, this pigment concentrated with increased temperature (Fig. 6). If the temperature response of the white pigment were thermoregulatory, then with increase in temperature the white pigment should have dispersed rather than concentrated in order to increase the area of the body surface that was able to reflect light and heat efficiently. Since heat and bright light are usually associated in nature, as in sunlight, the antagonistic responses to light and temperature may be a mechanism to maintain a steady state of the white pigment in spite of changes in temperature and intensity of illumination.

Melanin of the blue crab, *Callinectes sapidus*, also responds to illumination. Between 120 and 3000 ft c. the melanin became progressively more dispersed (Fingerman, 1956a). As the temperature increased from 10°–28°C, the melanin of *Callinectes* concentrated. The latter response can be interpreted as thermoregulatory. The tendency of melanin to concentrate with increased temperature, thereby reducing the light-absorbing area, may be a primitive attempt at homoiothermism.

The older literature on thermoregulatory use of melanophores by vertebrates is well known. At low temperatures melanin disperses so that more radiant energy is absorbed by the dark skin. This pigment then concentrates as the body temperature rises, the same situation as described above for the grasshopper. Further support of the thermoregulatory concept of chromatophores was offered by Deanin and Steggerda (1948). They demonstrated spectrophotometrically that more light is reflected from the skin of a frog with concentrated black pigment as a result of adaptation to a white background than from skin with dispersed melanin as a result of having been kept on a black background. These investigators also showed that more light of wavelengths from the red end of the spectrum than from the violet end is reflected from the skin of frogs on black and on white backgrounds. This difference seems significant in view of the fact that red rays have a greater heating capacity than violet rays. Edgren (1954a) found that the tree frog, *Hyla versicolor*, was dark at low temperatures, 3°–5°C, and lightened with increased temperature.

Secondary responses to light have been described among a wide variety of animals. Some species have the ability to mimic colored backgrounds as well as shades of gray, thereby demonstrating an ability to discriminate colors independent of intensity. In this manner Kühn (1950) showed that the cephalopods *Sepia officinalis* and *Octopus vulgaris* possess a true color vision.

Brown and Sandeen (1948) showed that in the fiddler crab, *Uca pugilator*, an albedo response operates to concentrate the melanin when specimens are on a white background and disperse it on a black background. However, because of a strong melanin-dispersing response to total illumination the melanin remains more concentrated in specimens on a black background than on a white one. This species also shows a 24 hr rhythm of color change, manifested by concentration of black and white pigments at night and dispersion during the day. Furthermore, this rhythm is the primary determinant of the coloration of the fiddler crab. The albedo response and the response to total illumination do not produce enough background adaptation to result in sufficient obliterative coloration that could have survival value. Brown (1950) found that the red pigment of this same crab exhibited extensive responses to background and is, therefore, better adapted for alteration of the body color in accordance with the background than are the black and white pigments. The red pigment dispersed in specimens on a black background and concentrated on a white background.

The black and the red chromatophores of the blue crab, *Callinectes sapidus*, showed a background response (Fingerman, 1956a). Both pigments were more concentrated in crabs in a white pan than in a black container. The blue crab also has a 24 hr rhythm of color change; both pigments are more dispersed by day than by night. In contrast to the situation observed in the fiddler crab, in the blue crab the albedo response and 24 hr rhythm contribute approximately equally to the coloration of the organism. Furthermore, in the blue crab the extent of black pigment migration observed when crabs were changed from black to white backgrounds and back again was greater than that observed with the red pigment, the reverse of the situation in *Uca pugilator*.

Dispersion of white pigment occurs when specimens of *Palaemonetes pugio* and *P. paludosus* are placed on a white background; concentration occurring on a black one (Fingerman, and Tinkle, 1956).

Fingerman (1957a) showed that in the dwarf crayfish, *Cambarellus shufeldti*, the red pigment dispersed maximally and the white pigment concentrated maximally when the specimens were placed on a black background at $22°-28°C$. The pigmentary states reversed themselves when specimens were on a white background. The white pigment of the crayfish, *Orconectes clypeatus*, showed the same background responses as that of *Cambarellus*, but the red pigment behaved differently (Fingerman, 1958). The red pigment of *Orconectes* would disperse maximally when specimens were put in a black container but only concentrated to an intermediate stage when *Orconectes* were on a white background. Maximal red pigment concentration did not occur in specimens kept on a white background for 32 days.

The melanophores of the grapsoid crab *Sesarma reticulatum* show an albedo response; the melanin is more dispersed in specimens on a black background than on a white one (Fingerman, Nagabhushanam and Philpott, 1960a, 1961). A primary response was also apparent, the degree of pigment dispersion increased with increase in incident light intensity (Fig. 7). The red chromatophores of the anomuran mud shrimp, *Upogebia affinis*, did not respond to background but this pigment concentrated with increase in incident light intensity (Fingerman, Nagabhushanam, and Philpott, (1960b).

In vertebrates secondary color changes characterize the later larval stages and the entire adult period. The organisms are typically dark on a dark background and light on a white one. The most striking secondary responses of all the vertebrates are shown by fishes. For example, in *Ericymba buccata* the average diameter of the melanin masses varies in a directly proportional fashion with the ratio of incident to reflected light (Brown, 1936). Strong

background responses have also been observed in the elasmobranch, *Scyllium canicula,* by Waring (1938). Breder and Rasquin
(1955) found that the angel fish, *Chaetodipterus faber,* showed
both intensity and albedo responses. The fish was solid black in

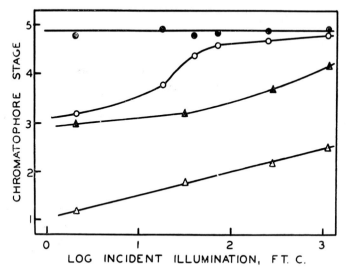

FIG. 7. Relationships between melanophore stage and the logarithm of
the incident light intensity for crabs in the day phase of the rhythm of
color change and on a black background (dots), day phase and on a
white background (circles), night phase and on a black background (solid
triangles), night phase and on a white background (empty triangles).
(From Fingerman, Nagabhushanam, and Philpott, 1961.)

bright light or against a light background and was banded black
and white in low light or against a dark or mottled background. This
response is interesting because of the fact that the fish is overall
darker when against a light background than when placed on
a dark background. The authors postulated this behavior has
adaptive significance in that the animal is banded on a mottled
background but uniform in color so as to appear similar to a bit
of sea bottom litter when on a light background.

Adaptation to background in both vertebrates and invertebrates
depends on a spatial separation within the retina of receptor

elements, stimulation of which produces either a lightening or darkening response (e.g. Sumner, 1933; Hogben and Slome, 1936; Smith, 1938; Hogben and Landgrebe, 1940). Sumner (1933) found that darkening the lower half of the field of vision by means of false corneas resulted in darkening of the fish *Fundulus parvipinnis*. Butcher and Adelmann (1937) performed experiments similar to those of Sumner using *Fundulus heteroclitus*. When the dorsal portion of the retina was illuminated, the fish became pale. When the ventral portion of the retina but not the dorsal portion was illuminated, the fish darkened. These observations were confirmed with specimens whose eyes had been rotated 180° by surgery. Their conclusion that the dorsal and ventral portions of the retina are physiologically different agrees with the findings of Sumner, even though these investigators disagreed on which portion of the retina is more important for the darkening process. Hogben and Slome (1936) also postulated that background responses in the amphibian *Xenopus laevis* depend on distinct localized retinal elements. Smith (1938) showed, on the basis of experiments in which different portions of the eyes of the isopod *Ligia oceanica* were covered with an opaque material or stimulated differentially, that stimulation of the dorsal portion of the retina resulted in melanin dispersion and stimulation of the ventral and lateral portions resulted in blanching. Essentially the same conclusions were arrived at by Hogben and Landgrebe (1940) for the stickleback, *Gasterosteus aculeatus*. Photoreceptors concerned with the black background response were in the floor of the retina below the optic nerve; those associated with the white background response were in the center of the retina above and below the blind spot. Presumably, in all animals investigated, except for *Fundulus parvipinnis* (Sumner, 1933), the upper portion of the retina of specimens on a white background is stimulated by reflected light much more than it is when specimens are on a black background and consequently, the lightening response is called forth.

Color changes associated with mating and parental behavior have not received much attention. Hadley (1929) observed that male lizards, *Anolis porcatus*, show a striking change from green to brown while pairing with a female. The significance of this color change is unknown. It may be used to drive off other males. Kramer (1960) observed that breeding pairs of the cichlid fish *Aequidans latifrons* exhibit striking color changes correlated with parental behavior. The vertical band chromatophores of the female become conspicuously black several days before egg laying. This coloration persists during egg laying and afterwards during egg fanning. Whenever a male relieves a female guarding and fanning eggs, his vertical banding becomes darker, then lessens or disappears as he leaves the eggs. When the fry begin to swim and the male participates with the female in guarding and schooling the fry, his banding intensity approximates the female's. Most intense banding appears in both parents when they confine the fry to a small territory, aggressively defending against other adults in the same tank.

METHODS OF STAGING CHROMATOPHORES

The shade of an organism depends on (1) the number of chromatophores, (2) the nature of the pigment each chromatophore contains, and (3) the degree of dispersion of the contained pigment. Early workers simply described the shade of an organism in macroscopic terms such as dark, intermediate, or light. Obviously, for quantitative work other systems had to be found. Several have been tried. These methods have been adequately described by Parker (1948). The most popular is that of Hogben and Slome (1931) who divided the entire range from maximum concentration to maximum dispersion into five stages (Fig. 8). According to their scheme, stage **1** represents the most concen-

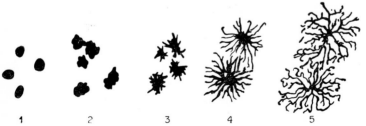

FIG. 8. Chromatophore stages of Hogben and Slome (1931.)

trated condition of the pigment, stage 5 the most dispersed, and stages 2, 3, and 4 the intermediate conditions. These five stages closely correspond to the descriptive stages of (1) punctate, (2) punctate-stellate, (3) stellate, (4) stellate-reticulate, and (5) reticulate. The obvious advantages of this system are that first it allows direct, accurate observation of individual chromatophores and second, it facilitates quantitative study and graphic representation of the changes of the degree of pigment dispersion. The major criticisms of the Hogben and Slome method are that (1) the method is subjective and (2) the stages may have no mathematical significance and might be construed as indicating, for example, that stage 4 represents twice as much pigment dispersion as stage 2.

The photoelectric method of Hill, Parkinson, and Solandt (1935), measurement of the fraction of incident light reflected from a unit area of skin surface, was devised to eliminate the subjective aspect of the method of Hogben and Slome. The obvious defect in the photoelectric method is that the amount of reflected light depends on both the degree of pigment dispersion and the number of chromatophores. Furthermore, in the same piece of skin one pigment may disperse and another concentrate, but the photoelectric method reveals only the net change in light reflected by the pigmented surface and not the changes in the individual chromatophores. Modifications of the original photoelectric technique have been used by Wright (1948, 1954a), Thing (1952), Rigler and Holzbauer (1953), Shizume, Lerner, and Fitzpatrick (1954), Deutsch, Angelakos, and Loew (1957), and Teague and Patton (1960). An advantage of the photoelectric technique is that dosage-response curves can be obtained that are mathematically true, i.e. the readings obtained by means of a photocell are directly proportional to the amounts of reflected or transmitted light.

Thing (1952) discussed the difficulties inherent in both techniques. As a result of his experiments he concluded that the results obtained by the two methods are nearly the same and that both methods are equally suitable for assays. He preferred the Hogben and Slome method because of its inherent simplicity.

CHAPTER 2

CHROMATOPHORES OF CRUSTACEANS

THE spectacular color displays exhibited by crustaceans have attracted the interest of biologists since the middle of the nineteenth century. In 1842 Kröyer reported the results of his observations of the prawn *Hippolyte*. Early investigators believed that pigment migration in crustaceans was mediated by nerves. The researches of Koller and Perkins were instrumental in dispelling this erroneous notion. Koller (1925, 1927) presented the first evidence that bloodborne substances were involved. He transfused blood from a dark specimen of the shrimp *Crangon vulgaris* maintained on a black background into a light specimen kept on a white background and noted that this light animal darkened. The reciprocal transfusion yielded negative results. Blood from an animal on a yellow background caused a yellowing of animals on a white background. Perkins (1928) found that denervation of a portion of the skin of the prawn *Palaemonetes vulgaris* had no effect on the chromatophores. He also observed that removal of the eyestalks resulted in permanent maximal dispersion of the pigment in the red chromatophores. Injection of extracts of eyestalks produced a transitory blanching of eyestalkless *Palaemonetes*. Since these early efforts, the chromatophore systems of several crustaceans have been investigated in detail, primarily to determine the sources and actions of the hormones responsible for migration of chromatophoral pigments.

SOURCES OF CHROMATOPHOROTROPINS

An interesting contrast between the development of vertebrate endocrinology and crustacean endocrinology is that in the former discipline the morphology of the endocrine glands was well estab-

19

lished long before the physiology of the structures was established whereas among the crustaceans the function of the endocrine sources was established prior to studies of their histology. Within the last decade striking advances have been made in the direction of describing the structure of secretory organs in crustaceans. Among the noteworthy accomplishments is the demonstration that substances produced in neurosecretory cells are transported along axons to storage and release centers which had previously been considered the actual sites of hormone synthesis.

After the initial discoveries of Perkins (1928) and Koller (1928) that chromatophorotropins are present in the eyestalk of crustaceans, systematic studies were undertaken to locate the sources of these hormones. Hanström (1933) found two structures in the eyestalk, the sinus gland and X-organ, which he suspected of having endocrine activity. In 1933 Brown supplied the first clear-cut evidence that central nervous organs outside the eyestalk contain chromatophorotropins. He used the prawn *Palaemonetes vulgaris*.

The sinus gland was found to contain a greater concentration of chromatophorotropin than any other structure in the eyestalk. Carlson (1936) found that the vast majority of melanin-dispersing hormone in the eyestalk of the fiddler crab, *Uca pugilator*, was located in the portion of the eyestalk which contained the sinus gland. Brown (1940) arrived at the same conclusion for the red pigment-concentrating principle in the eyestalk of *Palaemonetes*.

In 1937 Hanström presented the results of a detailed, systematic study of the location and structure of the sinus gland in a large number of crustaceans. He later published three additional summaries of his studies (Hanström, 1939, 1948a, b). The most primitive sinus gland occurs in the Mysidacea (*Eucopia*) and Euphausiacea (*Meganyctiphanes*). In these organisms the sinus gland consists simply of a thickened disc-shaped portion of the epineurium that encloses the central nervous centers that have migrated into the eyestalk during its evolution. In the Decapoda (*Palaemonetes, Crangon, Palaemon, Systellaspis*) the sinus gland is a more or less beaker-shaped structure surrounding the opening of a small dorsolateral inner blood sinus into a larger outer blood sinus in the region between the medulla interna and the medulla externa. In crabs the inner dorsolateral sinus branches near its opening and the gland extends over the proximal portions of the branched sinus. In the Reptantia Astacura (*Astacus, Cambarus, Homarus*) the branching of the inner sinus is more pronounced than in the crabs with the result that the sinus gland is more diffuse. Some portions of the gland project into the outer sinus, other portions surround the proximal parts of the branched inner sinus opening. In several Anomura with reduced eyes the sinus gland occurs in the head rather than in the eyestalk. In these forms (*Emerita, Upogebia, Calocaris, Callianassa*) the structure of the gland has been secondarily simplified. The gland is connected to the epineurium of the supraesophageal ganglia and adjoins a large outer blood sinus, but it has no connection with an inner sinus. In the sessile-eyed Isopoda and other eyestalkless crustaceans the sinus gland lies in the head in the vicinity of the optic lobe, adjoining a blood sinus.

The X-organ was the only other structure in the eyestalk that Hanström thought could possibly be a source of hormone. He postulated that the X-organ has evolved from the sense cells of a paired frontal organ that sometimes occurs as an "eye-papilla" or is sometimes reduced to a "sensory pore".

In 1941 Welsh described a structure in the eyestalk of the crayfish *Cambarus bartoni* that he termed the X-organ. Hanström (1948b) had been unable to find his X-organ in the crayfish *Astacus vulgaris*. The X-organ described by Welsh lay in close proximity to the medulla terminalis and consequently was not the X-organ described by Hanström in other crustaceans. In 1953 Carlisle and Passano helped to clarify the situation through a study of the morphological relationships of the two "X-organs" in the eyestalks of 11 species of crustaceans. In the Natantia investigated, the X-organs of Hanström and Welsh are widely separated but connected by a nerve. In the Stomatopoda (*Squilla mantis*) the two organs are closer together than in the Natantia and still connected by a nerve. In the Brachyura, however, where the sensory pore is poorly developed or lacking, both organs form a single complex usually at the proximo-ventral corner of the medulla terminalis. Carlisle and Passano suggested that when the two portions of the gland are separated, the X-organ of Hanström should be called the pars distalis X-organi and that of Welsh the pars ganglionaris X-organi. However, in 1956 Knowles and Carlisle decided that the former organ should be known as the Sensory Pore X-organ or Sensory Papilla X-organ and the latter the Medulla Terminalis X-organ. In the opinion of the present author such name changes lead only to confusion on the part of readers of the literature.

Hanström (1948b) stated that the secretory products of the X-organ are in part small droplets of an eosinophilic and fuchsinophilic substance and in part larger irregularly shaped concretions of a concentric structure. Carlisle (1953a) believes that the latter concretions are actually terminations of axon fibers each of which forms a many-layered club-shaped body. He (Carlisle, 1953b) referred to these as "onion bodies" because of a general resemblance when seen in cross section to the ensheathing scales of an onion.

Several investigators in the early 1950's began working on different phases of crustacean endocrinology with specimens whose sinus glands alone had been removed from the eyestalks (Knowles, 1950, 1952; Bliss, 1951; Frost, Saloun, and Kleinholz, 1951; Havel and Kleinholz, 1951; Passano, 1951a, b; Travis, 1951; Welsh, 1951). The effects of sinus gland ablation were not the same as those observed after eyestalk removal. The evidence suggested that the sinus gland is merely a storage and release center for neurosecretory material produced elsewhere. This concept is the only tenable explanation of studies of molting in eyestalkless and sinus glandless fiddler crabs (Passano, 1953). Likewise, what is known about chromatophorotropins in the eyestalk falls in line with this theory.

After these original studies the neurosecretory systems of several crustaceans were mapped and attempts were made to correlate cytological signs of neurosecretory activity with function. Carlisle (1953c) has indicated that in *Lysmata seticaudata* the onion body components of the sensory papilla X-organ are the terminations of axon fibers some of which originate in cell bodies which lie in the medulla terminalis and others which appear to originate in the supraesophageal ganglia. It was suggested that substances are produced in the cell bodies of the axons and transported along the axons to the onion bodies where they are stored preparatory to release.

The sinus gland seems to receive axons from several groups of neurosecretory cells. Cells in the medulla terminalis appear to be the largest contributors to the sinus gland (Enami, 1951a). Enami noted that secretory droplets could be detected along the path of axons arising from cells with secretory activity.

Bliss and Welsh (1952) studied the anatomical relationships between neurosecretory centers in the eyestalks and supraesophageal ganglia of the land crab *Gecarcinus lateralis*. These investigators found that the sinus gland in this crab is a mass of swollen nerve fiber endings, arranged in the form of an inflorescence and bearing secretory material. The histological structure of regenerated sinus glands, which appear after sinus gland removal, resembles that of normal sinus glands. The nerve fibers whose endings compose the sinus gland originate in the neurosecretory cells of the supraesophageal ganglia, the eyestalk ganglia, medulla terminalis X-organ, and possibly the thoracic nerve cord.

One site of a release center for chromatophorotropins outside the eyestalk was indicated by the investigation of Brown and Ederstrom (1940) who found that the portion of the circumesophageal connectives just posterior to the connective ganglia contained more melanin-dispersing hormone for *Crangon septemspinosus* than other central nervous organs outside the eyestalk. Brown (1946) later found that the tritocerebral commissure which runs posteriorly from the connective ganglion on one side down the medial surface of the connective for a short distance and then proceeds to the contralateral connective around the posterior side of the esophagus, together with its immediate continuation within the circumesophageal connectives, contained all of the *Crangon*-darkening hormone (CDH) occurring within the central nervous system. Brown, Webb, and Sandeen (1952) found that the trito-cerebral commissure was a potent source of red pigment concentrating hormone in the prawn *Palaemonetes vulgaris*. The tritocerebral commissure contained as much of this hormone as did one sinus gland.

A detailed morphological and physiological study of the tritocerebral region in the decapods *Penaeus brasiliensis* and *Palaemon serratus* was published by Knowles (1953). Two fine nerves leave the tritocerebral commissure and supply paired muscles nearby. Many acidophilic neurosecretory droplets lie along the course of these nerves. In *Penaeus brasiliensis* at the point where each post-commissure nerve reaches the muscle which it innervates there is an enlargement of the epineurium which forms a flat plate filled with acidophilic droplets; this is partially fused to the wall of a blood sinus. This "sinus-plate" seems in many respects to be analogous to the sinus gland of the eyestalk. In *Palaemon serratus* no discrete sinus-plate was seen, but the post-commissure nerves were wide and flat for the proximal part of their course; many acidophilic droplets lay in the flat portion of the post-commissure nerves and in the epineurium of the commissure nearby. The droplets in *Palaemon serratus* lay along definite pathways which appeared to radiate from the bases of the post-commissure nerves. Large cells suspected to be secretory were found within the commissure, close to its hinder margin near the points of emergence of the post-commissure nerves and in the connective ganglia. Fibers from these latter cells pass along the post-commissure nerves. Injection experiments revealed that those regions richest in acidophilic droplets yielded extracts that contained large quantities of red and white pigment-concentrating principles. Knowles suggested that these chromatophorotropins are produced by neurosecretory cells in the tritocerebral commissure and in the connective ganglia, and are stored and released by specialized portions of the epineurium of the post-commissure nerves and of the commissure. These specialized portions have been termed the post-commissure organs (Carlisle and Knowles, 1959). Maynard (1961), the first investigator to describe the post-com-

missure organs of a brachyuran, found that these organs in *Pachygrapsus crassipes* are very similar to those of *Penaeus* and *Palaemon*. One of the earliest pieces of evidence in support of the concept that chromatophorotropins occur in central nervous organs outside the eyestalk was the observation that stimulation of the eyestalk stubs of the prawn, *Palaemonetes*, results in transitory blanching whereas the fiddler crab, *Uca*, darkens (Brown, 1939b, 1952b). These chromatophorotropins are presumably released from the post-commissure organs which possess all the characteristics of neurohaemal organs. Carlisle and Knowles (1953) defined a neurohaemal organ as one which consists primarily of the terminations of neurosecretory fibers and is located in or alongside the wall of blood vessels or blood cavities.

Matsumoto has published an interesting series of papers on neurosecretory cells in the thoracic ganglion of *Eriocheir japonicus*. He (Matsumoto, 1954a) found three types of neurosecretory cells in the thoracic ganglion of *Eriocheir japonicus*. In contrast, Enami (1951a) observed only one type of neurosecretory cell in the thoracic ganglion of *Sesarma*. In another investigation, Matsumoto (1954b) correlated the presence of chromatophorotropin with the distribution of one type of neurosecretory cell in the thoracic ganglion of *Eriocheir*. This neurosecretory cell, type B as he called it, appears to be the main source of melanin-dispersing hormone in the thoracic ganglion. The effectiveness of an extract on the melanophores could be predicted on the basis of the number of type B cells in the extracted portion of tissue. Matsumoto (1956) studied neurosecretion in another crab, *Chionoecetes opilio*. By tracing the neurosecretory axons it became clear to him that a large part of the secretory material flowed into the circumesophageal connectives toward the supraesophageal ganglia and the eyestalks. However, some of the neurosecretory axons entered the first, second, third, and fourth pedal nerves but the ultimate fate of these axons was not clear. Some of the neurosecretory material was stored also in tissue spaces in the mid-dorsal and latero-caudal portions of the thoracic ganglion. In 1958 Matsumoto presented the results of a lengthy and detailed comparative study of neurosecretory cells in five species of crabs, *Potamon* (*Geothelphusa*) *dehaani, Eriocheir japonicus, Chionoecetes opilio, Neptunus* (*Neptunus*) *trituberculatus,* and *Sesarma* (*Sesarma*) *intermedia.* Eleven different types of neurosecretory cells were observed. He studied the morphogenesis

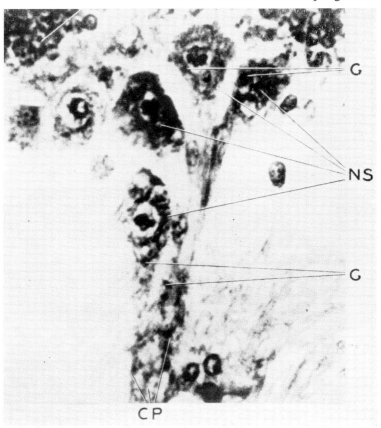

FIG. 9. Neurosecretory cells (NS) of the eyestalk of *Gecarcinus lateralis.* Secretory granules, G; cell processes, CP. (From Bliss, Durand, and Welsh, 1954.)

of neurosecretory cells in *Potamon.* Secretory granules first appeared one or two days prior to hatching. The embryonic sinus gland, composed of nerve endings, as in the adult contained neurosecretory products only when the cell type he had earlier

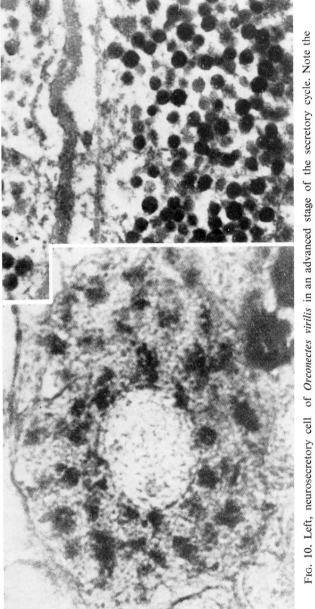

FIG. 10. Left, neurosecretory cell of *Orconectes virilis* in an advanced stage of the secretory cycle. Note the aggregations of granules. Cells of this type are numerous in the X-organ. ×2,100. (From Durand, 1956). Right portions of two axon terminals in the sinus gland of *Cambarellus shufeldti* showing neurosecretory granules observed with an electron microscope. ×64,000. (From Fingerman and Aoto, 1959.)

termed "B" began to secrete. After eyestalk severence, accumulation of neurosecretory products occurred near the cut end, indicative of axonal transport. Four kinds of neurosecretory cells were found in the central nervous organs of the isopod *Armadillidium vulgare* by Matsumoto (1959). These cells were similar to some of the neurosecretory cells he had previously found in crabs.

Bliss, Durand, and Welsh (1954) compared the neurosecretory systems of the land crab *Gecarcinus lateralis* and the crayfish *Orconectes virilis*. Morphologically the neurosecretory systems were similar (Fig. 9). Durand (1956) reported that there are four cytologically distinct types of neurosecretory cells in the eyestalk and supraesophageal ganglia of *Orconectes virilis*. Two of these types are restricted to the medulla terminalis X-organ (Fig. 10). The other two cell types occur in all neurosecretory cell groups in the supraesophageal ganglia and eyestalk except the X-organ.

Fingerman and Aoto (1959) mapped the distribution of neurosecretory cells in the supraesophageal ganglia of the dwarf crayfish, *Cambarellus shufeldti* (Fig. 11). Five groups of neurosecretory cells were found in the supraesophageal ganglia. The medulla terminalis X-organ appeared to contain two types of neurosecretory cells just as was found in *Orconectes virilis* by Durand (1956). The sinus gland of *Cambarellus* also appeared to consist of nerve endings swollen with neurosecretory material. The blood sinus adjacent to the sinus gland contained large numbers of secretory droplets.

Bliss, Durand, and Welsh (1954) and Durand (1956) presented diagrams depicting the distribution of groups of neurosecretory cells in the supraesophageal ganglia of *Orconectes virilis*. The description reported by Durand (1956) differs somewhat from the earlier description (Bliss, Durand, and Welsh, 1954). Five major groups of neurosecretory cells were reported at first, but Durand (1956) believed that two of these were actually one. Also in the original report a group (B_3 in the terminology of Bliss, Durand, and Welsh, 1954) was reported to lie lateral to the olfactory lobes. But in the later publication Durand stated it was actually located mesial to the olfactory lobes and on the lateral side of the mass of fibers in the supraesophageal ganglia. The distribution of neurosecretory cells in the supraesophageal ganglia of *Cambarellus* described by Fingerman and Aoto (1959) appears to be a composite of the two descriptions reported for *Orconectes*

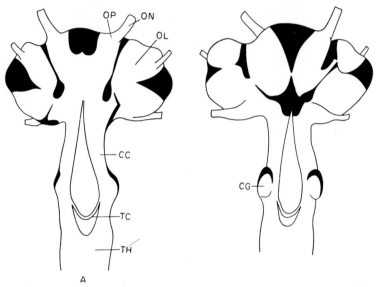

FIG. 11. Dorsal (posterior) view (A) and ventral (anterior) view (B) of the supraesophageal ganglia and circumesophaegal connectives of *Cambarellus* showing in black the regions of neurosecretory cells. CC, circumesophageal connective; CG, circumesophageal connective ganglion; OL, olfactory lobe; ON, optic nerve; OP, optic nerve peduncle; TC, tritocerebral commissure; TH, thoracic nerve cord. (From Fingerman and Aoto, 1959.)

virilis. The group of cells (B₃) that Durand decided is mesial rather than lateral to the olfactory lobes is indeed lateral in the dwarf crayfish. This group is by far the most obvious one in histological sections of the supraesophageal ganglia of *Camberellus* as well as in vitally stained preparations. A group that was termed B₂ in the earlier account but which was consolidated by Durand with the B₃ group is also obvious as a distinct group in the supraesophageal ganglia of *Cambarellus*. On the other hand, the distribution of cell group B_4 in the publication of Durand (1956) corresponds more closely with that of the two groups of cells in *Cambarellus* that lie near the origin of the circumesophageal connectives than does the distribution of B_4 cells in the publication of Bliss, Durand, and Welsh (1954).

Potter (1954, 1958) cited some extremely interesting results he obtained from studies of the medulla terminalis X-organ-sinus gland complex of the blue crab, *Callinectes sapidus*. He found at least six distinct tinctorial types of fine particles in the sinus gland, produced by six distinct types of neurosecretory cells in the medulla terminalis X-organ. The staining reactions of the fine particles are summarized in Table I.

An accessory sinus gland has been found in four species of isopods, *Idotea japonica*, *Idotea ochotensis*, *Cleantilla isopus*, and *Mesidotea* sp., by Oguro (1959a, b). The ordinary sinus gland is located near the middle portion of the optic lobe whereas the accessory sinus gland is situated at the distal portion of the optic lobe near the lamina ganglionaris (Fig. 12). Both sinus glands appear to be composed of bulbous endings of axons from neurosecretory cells, as has been described in decapod crustaceans. Some differences were apparent between the two sinus glands. Both glands were usually filled with acidophilic granules but portions of the ordinary sinus gland occasionally had basophilic granules whereas the accessory sinus gland never exhibited a basophilic tendency.

Miyawaki has also been interested in neurosecretory cells. He (1956) reported on his studies of the supraesophageal ganglia and thoracic ganglia of the crab *Telmessus cheiragonus*. Neurosecretory cells were stained according to the periodic acid-Schiff (PAS) method. The neurosecretory material appeared to be PAS-positive, thereby suggesting that neurosecretory granules contain polysaccharide. Miyawaki (1960a) later found some interesting globules in the cytoplasm of neurosecretory cells in six species of decapod crustaceans, among which were *Telmessus cheiragonus* and *Gaetice depressus*. These globules are variable in size, ring-

TABLE 1

NEUROSECRETORY CELLS OF THE X-ORGAN-SINUS GLAND COMPLEX IN BOUIN-FIXED EYESTALKS OF ADULT *Callinectes sapidus* (From Potter, 1958).

Cell type	Dye which stains the fine particles	Approximate cell diameter μ	Approximate number of cells	Approximate % of endings in the sinus gland
I	azocarmine (red)	50	100	females: 60 males: 75
II	orange G (yellow)	50	20	10
III	azocarmine and aniline blue (purple)	60	females: 60 males: 2 ?	females: 20-25 males: 1 ?
IV	aldehyde fuchsin (purple)	15	25	5
V	orange G (orange)	20	females: 8 males: 11	5-10
VI	aniline blue (blue)	30	6	5-10

like or crescents, and stain darker at their periphery than in their center. They may be the site of elaboration of neurosecretory products. A weak point in his argument is that he found these globules in all neurons whether or not they were neurosecretory. The same author (Miyawaki, 1960b, c) showed by histochemical techniques that these cytoplasmic globules are composed of phospholipids and cerebrosides, and are embedded in or entangled with the Nissl granules which are composed of RNA.

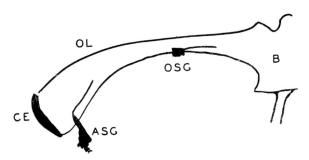

FIG. 12. Diagram showing the morphological relationship of the ordinary sinus gland (OSG) to the accessory sinus gland (ASG) in the isopod *Idotea japonica*. B, brain; OL, optic lobe; CE, compound eye. (Redrawn from Oguro, 1959a.)

The electron microscope has recently been utilized to study neurosecretory cells and their products in crustaceans. Knowles (1958) described the results of his observations of the post-commissure organ of the mantid shrimp, *Squilla mantis*. Mitochondria in the neurosecretory cells of this organ appeared to be at least five times more numerous and also larger than those in motor fibers of neurons that were not neurosecretory. The high incidence of mitochondria in the neurosecretory fibers indicates that considerable biochemical activity occurs there, but it is impossible to tell whether these changes are involved in the manufacture of secretory materials or whether they energize the passage of secretory material along the axons; possibly they may be actively engaged both in the production of material and in its transport. When the neurosecretory fibers reach the site of release, some of them form loops just beneath the surface of blood sinuses, possibly to provide regions for storage of material. The neurosecretory fibers emerge from the sheath of Schwann which had enveloped them to terminate in swellings approximately 1 μ in diameter.

The sinus gland of the land crab *Gecarcinus lateralis* was observed by means of the electron microscope by Hodge and Chapman (1958). The dilated axon endings of the sinus gland were filled with dense granules that appeared to be bounded by a delicate membrane. In the sinus gland two size ranges of granules were noted, 0.05–0.10 μ and 0.15–0.20 μ. Each axon ending, however, contained granules of a nearly uniform size. Endings with the larger size granules predominated. The axons in the sinus gland are nonmyelinated with thin limiting membranes and possess many neurofibrils.

Knowles (1959) agreed with Hodge and Chapman that neurosecretory granules are bounded by a thin membrane. He extended his observations to the sinus gland of *Squilla mantis* and the postcommissure organs of *Palaemon serratus*. Each neurosecretory droplet appeared to be bounded by a multi-layered membrane that appeared to consist essentially of two electron-dense layers with a less dense layer between them; each layer measured approximately 30 Å. However, Knowles stated that the concept of discrete spherical or ovate droplets of secretory material each entirely surrounded by a multi-layered membrane does not provide a completely satisfying explanation of the observed facts because (1), within certain droplets still more membranes may be seen and (2), in certain sections the droplets are not spherical but extremely elongated and have the appearance of beads of secretory material lying in tubules within the cytoplasm. An hypothesis that the secretory material in neurosecretory fibers is contained in fine tubules, and not in spherical units helps to explain the variation in droplet size. This system of tubules in the cytoplasm of neurosecretory cells bears a striking resemblence to the endoplasmic reticulum system which is said to be characteristic of active secretory cells.

3

Another study of neurosecretory cells by means of the electron microscope was that of Fingerman and Aoto (1959). These investigators studied the sinus glands and central nervous organs of the dwarf crayfish, *Cambarellus shufeldti* (Fig. 10). The numerous mitochondria of neurosecretory cells were also noted by these investigators. The diameter of the neurosecretory granules in the dwarf crayfish ranged from 30–300 mμ. This size range agreed well with the sizes of granules reported by Hodge and Chapman (1958) and Knowles (1959) for crustaceans, Duncan (1955, 1956) for chickens and Palay (1955) for rats. Fingerman and Aoto also observed vesicles 3–8 μ long in some ganglionic cells of the circumesophageal connectives. The neurosecretory granules, in contrast to the observations of Hodge and Chapman (1958) and Knowles (1959), did not appear to be bounded by a distinct membrane. Palay (1955) had reported such a bounding membrane in the rat pituitary, but Duncan (1955, 1956) did not observe them in chickens.

Physiological evidence in favor of the hypothesis that chromatophorotropins occur within neurosecretory granules of crustaceans which are bounded by a semipermeable membrane was presented by Pérez-González (1957). She used homogenates of sinus glands from the fiddler crab, *Uca*. Melanin-dispersing hormone was liberated when homogenates were made hypotonic whereas homogenates prepared in isotonic sucrose caused little dispersion of melanin. The amount of melanin dispersion produced by extracts in distilled water was greater than that caused by preparations containing 10 times more sinus gland tissue extracted in isotonic sucrose (Fig. 13). Fingerman (1959b), using eyestalks and supraesophageal ganglia plus the circumesophageal connectives of *Cambarellus shufeldti*, did not find that hypotonic extracts were more potent than isotonic extracts. The results of Fingerman (1959b) could be anticipated since Fingerman and Aoto (1959) did not observe a membrane around the neurosecretory granules of *Cambarellus*.

In 1960 Fingerman and Aoto reported that removal of both eyestalks of the dwarf crayfish resulted in accumulation of neurosecretory material in the supraesophageal ganglia and in the stubs of the optic nerves. These stubs hypertrophied and became engorged with neurosecretory droplets. A portion of the accumulated material proved to be red pigment-concentrating hormone. These observations were consistent with the hypothesis that neurosecretory material produced outside the eyestalk is transported into the eyestalk via axons in the optic nerve.

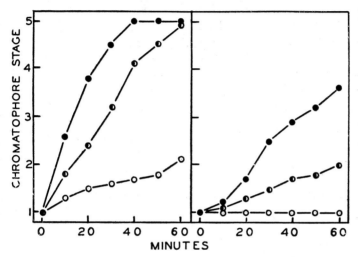

FIG. 13. Relationship between responses of melanophores of *Uca pugilator* to extracts of sinus glands in distilled water (left) and in 1.3M (isotonic) sucrose (right) and chromatophore stage. Circles, 0.002 sinus gland per ml; half-filled circles, 0.02 sinus gland per ml; dots, 0.2 sinus gland per ml. (Redrawn from Pérez-González, 1957.)

Many more studies of neurosecretory systems in crustaceans are necessary before a complete understanding will have been attained. The electron microscope reveals structures about which no one as yet can make an enlightened statement concerning their function. Much more remains to be done to determine the ultimate fate of the numerous neurosecretory products. Probably the largest problem is to correlate the distribution of the various types of neurosecretory cells and their secretory products with function.

3*

CHEMICAL NATURE OF CRUSTACEAN CHROMATOPHOROTROPINS

The chemistry of crustacean chromatophorotropins is an area about which very little concrete information can be given. Up to the present time not one crustacean chromatophorotropin has been sufficiently purified to allow adequate chemical analysis in spite of the fact that the problem has been under consideration since the early days of crustacean endocrinology.

Carlson (1936) was the first investigator who attempted to determine the chemical properties of eyestalk extracts. He found that the melanin-dispersing hormone of *Uca pugilator* would readily diffuse through a cellophane membrane. This hormone is soluble in ethyl alcohol but insoluble in ether and can be boiled for a few minutes in dilute HCl or NaOH without loss of activity. These results show the hormone has a relatively low molecular weight, is heat-stable, and is not a lipid. Abramowitz (1937a) described his early attempts to purify this same hormone by extraction in 95% ethyl alcohol. He concentrated the hormone about 3-fold but the overall loss of activity was 60%. Abramowitz and Abramowitz (1938) found that exposure of this hormone to NaOH for 3 min. had no effect, just as found by Carlson (1936). However, boiling in 0.1N NaOH for 10 min. resulted in complete loss of activity whereas boiling in 0.1N HCl for 30 min. did not affect the hormone and in fact appeared to increase the activity of the extract. Attempts to regenerate the activity of the material boiled in NaOH with an equivalent amount of HCl were unsuccessful. In 1940 Abramowitz, using adsorption techniques, succeeded in concentrating the melanin-dispersing hormone 100–200 times. The hormone reacted in a manner characteristic of amino bases.

Experiments performed in the fifties led Knowles and Carlisle (1956) to postulate that crustacean chromatophorotropins are peptides. The results of subsequent investigations lend support to this hypothesis. Carstam (1951) was the first one to study enzymatic breakdown of chromatophorotropins. He incubated extracts of the sinus glands of *Palaemon squilla* at 18°C for 3 hr with extracts of hypodermis and of hepato-pancreas and found the red pigment-concentrating hormone was destroyed. The material that caused the inactivation of the chromatophorotropin was heat-labile. Its action was proportional to its concentration and the time of action on the hormone. Consequently, he postulated that the hormone inactivating substance was an enzyme. However, he also reported that the proteolytic enzymes pepsin and trypsin had no effect on this hormone. This is surprising in view of the results of later investigators. Perhaps his sinus gland extracts contained something that inhibited the pepsin and trypsin. Knowles, Carlisle, and Dupont-Raabe (1956) found that the red pigment-concentrating hormone in the post-commissure organs of *Palaemon serratus* is destroyed by trypsin and by acid hydrolysis, indicative of peptide bonds. This substance was not inactivated by amine oxidase or by orthodiphenoloxidase, thereby suggesting that the red pigment-concentrating substance is not a catechol amine (Carlisle and Knowles, 1959).

Östlund and Fänge (1956) increased by means of column chromatography the titer of a substance in the eyestalks of the shrimp *Pandalus borealis* that concentrated the pigment in the small and the large red chromatophores of *Palaemon squilla*. They found this hormone was soluble in acetone whereas Abramowitz and Abramowitz (1938) reported that the black pigment-dispersing hormone of *Uca* was insoluble in acetone. This difference suggests that the two substances are quite different molecules. Östlund and Fänge postulated that the hormone might be a simple aromatic amine. However, they have revised their original opinion and now feel it may be a polypeptide (Edman, Fänge, and Östlund, 1958). The latter three investigators continued the earlier efforts to purify the red pigment-concentrating hormone from the eyestalk of *Pandalus borealis*. Beginning with several kilograms—they did not state the exact quantity—they ended with 30–40 mg of a slightly brownish, somewhat sticky substance that was still far from pure. The final product contained approximately 38 times more activity than an equivalent weight of fresh eyestalk. The hormone was readily soluble in ethanol and other low molecular weight alcohols but nearly insoluble in non-polar organic compounds and was inactivated by trypsin and chymotrypsin. Heating for 6 hr at 100°C in 6N HCl completely destroyed the hormone. The electrophoretic behavior of this hormone showed an acid character. From pH 2.5–9.0 it was found that the hormone migrated slightly toward the anode, the

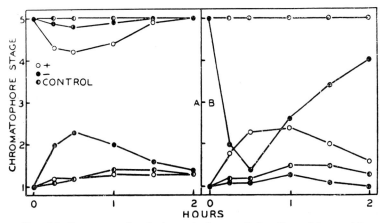

FIG. 14. Responses of red chromatophores of dwarf crayfish on white and on black backgrounds to extracts of eyestalks (A) and supraesophageal ganglia with the circumesophageal connectives attached (B). Filter paper electrophoresis was carried out at pH 7.5 on the extracts before injection. Circles, fraction that migrated toward positive pole; dots, fraction that migrated toward negative pole; half-filled circles, control. (From Fingerman and Aoto, 1958a.)

movements being somewhat more pronounced at high pH values than at low pH values. Finally, no chemical reaction was found by which the hormone could be identified. The ninhydrin test and the peptide reaction of Reindel and Hoppe (1954) were negative.

Pérez-González (1957) had found that the melanin-dispersing hormone from the sinus glands of *Uca pugilator* is inactivated by chymotrypsin as well as by extracts of hepato-pancreas. Fingerman (1959b), using trypsin, incubated extracts of eyestalks and supraesophageal ganglia plus circumesophageal connectives of *Cambarellus shufeldti* at 33°C for an hour and found almost complete inactivation of the red pigment-concentrating and red pigment-dispersing hormones. In spite of the growing feeling that chromatophorotropins of crustaceans are polypeptides, one must be cautioned that much of this sentiment is based on inactivation by proteolytic enzymes whereas trypsin and chymotrypsin, for example, also have a weak esterase activity and inactivation by these enzymes is in itself insufficient proof of the peptide nature of these chromatophorotropins.

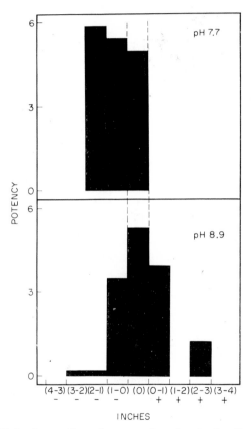

Fig. 15. Red pigment-dispersing potencies of extracts of eyestalks of *Cambarellus* versus position on the filter paper strip after electrophoresis at pH 7.7 and 8.9 (From Fingerman and Mobberly, 1960a.)

Fingerman and Aoto (1958a) found that the red pigment-dispersing and concentrating principles in the eyestalk of *Cambarellus* are different from the red pigment-dispersing and concentrating hormones of the supraesophageal ganglia and circumesophageal connectives (Fig. 14). At the same pH the substances in the eyestalk migrate in an electric field in opposite directions

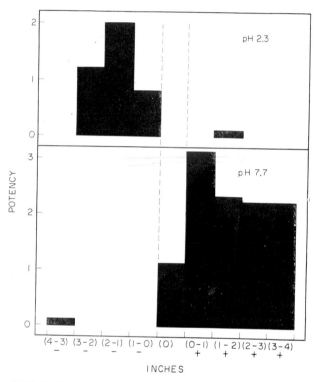

FIG. 16. Red pigment-dispersing potencies of extracts of supraesophageal ganglia with the circumesophageal connectives attached of *Cambarellus* versus position on the filter paper strip after electrophoresis at pH 2.3 and 7.7. (From Fingerman and Mobberly, 1960a.)

from their functional counterparts in the supraesophageal ganglia and circumesophageal connectives. Fingerman and Mobberly (1960a) studied the electrophoretic behavior of the red pigment-dispersing substances at different pH values. The substance from the eyestalk is electropositive at pH 7.7. Its isoelectric point appears to be near pH 8.9. At pH 2.3 the principle from the supraesophageal ganglia and circumesophageal connectives is electropositive but electronegative at pH 7.7. Its isoelectric point is probably near pH 5.0 (Figs. 15, 16). These data lend support to the hypothesis that crustacean chromatophorotropins are peptides because of the charge reversal these substances showed at different pH values.

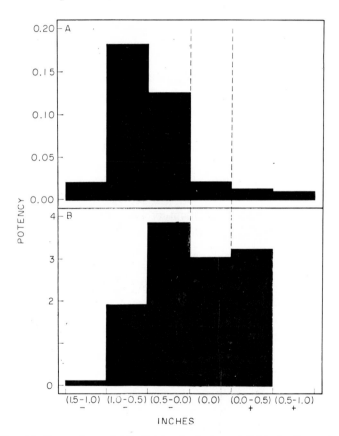

FIG. 17. Comparison of (A) distal retinal pigment light-adapting potencies and (B) red pigment-dispersing potencies of eyestalk extracts of dwarf crayfish subjected to filter paper electrophoresis at pH 9.0 for 2 hr. (From Fingerman and Mobberly, 1960a.)

It has been assumed that the distal retinal pigment light-adapting hormone of the eyestalk is a substance distinct from the chromatophorotropins of the eyestalk, but no one had firmly established this point (Carlisle and Knowles, 1959). Fingerman and Mobberly (1960a, b) were able to demonstrate that the distal retinal pigment light-adapting hormone, red pigment-concentrating hormone, and red pigment-dispersing hormone in the eyestalk of the dwarf crayfish, *Cambarellus shufeldti*, must be three distinct substances. Filter paper electrophoresis revealed that at pH 7.5–7.8 the pigment-dispersing hormone and the light-adapting substance were electropositive whereas the pigment-concentrating hormone was electronegative and, therefore, must be different from the retinal pigment activator. In this pH range the pigment-dispersing hormone and the light-adapting substances were electrophoretically indistinguishable. However, when electrophoresis was performed at pH 9.0 the charge on some of the molecules of red pigment-dispersing hormone reversed with the result that some of this substance was found on the anodal portion of the filter paper strip. However, no significant quantity of light-adapting material had become electronegative (Fig. 17). The only logical conclusion, therefore, is that the light-adapting hormone is different from both chromatophorotropins in the eyestalk.

CHROMATOPHORES OF ISOPODS

The fact that isopods have functional melanophores has been known for many years. In several species background responses and 24 hr rhythms of pigment migration, evidenced by melanin dispersion during the day-time and concentration at night, have been observed. Melanophores, leucophores, and xanthophores occur in this group of crustaceans.

Kleinholz (1937) published the first account of a study of the endocrine regulation of isopod chromatophores. He reported that color changes in *Ligia baudiniana* were due primarily to the melanophores. The animals were darker on a black background than on a white one (Fig. 18). Specimens kept in darkness exhibited a 24 hr rhythm of color change. Injection of aqueous extracts of heads into dark specimens brought about concentration of the pigment in melanophores.

Ståhl (1938a, b) reported that extracts of heads from *Oniscus asellus, Porcellio scaber,* and *Saduria entomon* caused dispersion of red and yellow pigment in pale *Palaemon squilla.* No response was observed when these extracts were injected into eyestalkless *Palaemon.* Extracts of heads of *Idotea balthica* injected into eyestalkless *Palaemon* produced no pigment dispersion but a very small amount of pigment concentration. The lack of a strong effect with the latter extracts may have been due to the fact that he filtered

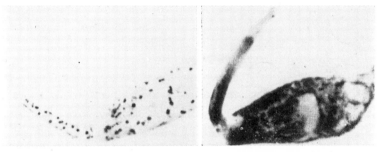

FIG. 18. Portions of appendages from light and dark *Ligia baudiniana.* (From Kleinholz, 1937.)

the extracts and the chromatophorotropin probably became adsorbed to the filter paper. McVay (1942) reported a great loss of chromatophorotropic activity when extracts of central nervous organs from the crayfish *Orconectes immunis* were filtered. Smith (1938) postulated the existence of pigment-dispersing and pigment-concentrating substances for the melanophores of *Ligia oceanica* on the basis of studies of the time relations of background adaptation.

Enami (1941a) found that another isopod, *Ligia exotica,* also shows striking background changes as well as the usual 24 hr rhythm of color change that persisted in specimens in constant darkness but whose expression was immediately inhibited in specimens under constant illumination on black and on white backgrounds. Enami (1941b) found that extracts of heads from *L. exotica* caused a distinct dispersion of the pigment in melanophores of light specimens. When the same extract was injected into dark animals having maximally dispersed pigment, a slight transitory blanching was noticed. This blanching may have been due to a black pigment-concentrating hormone that was able to express itself only briefly before being dominated by the potent black pigment-dispersing substance whose effect was so obvious. Nagano (1949) also studied the pigmentary system of *L. exotica* and found, in contrast to the results of Enami (1941b), that the response to extracts of heads was pigment concentration alone; no dispersion was observed, Fingerman (1956b) reinvestigated the chromatophore system of *L. exotica* because of the conflicting results of Enami (1941b) and Nagano (1949). The results of Fingerman agreed closely with those of Enami. Fingerman injected extracts of the sinus glands, cerebral ganglia, and thoracic nerve cord, and observed pigment dispersion alone. The rapid background responses and daily rhythm of color change observed by Enami (1941a) were also confirmed by Fingerman. The cycle reported by Fingerman was observed in animals maintained on black and on white backgrounds under an illumination of 40 ft c. and in darkness. However, the amplitude of the rhythm of the animals on an illuminated white background was greatly reduced as compared with the two other groups of *Ligia*. The amplitude was greatest in animals in darkness. At all times of day and night the pigment in melanophores was more dispersed in specimens on a black background than on a white one. These results are at variance with the findings of Enami (1941a) who reported that light inhibited the expression of this rhythm in the same animal.

Okay (1943) reported that the melanophores of *Sphaeroma serratum* respond to black and to white backgrounds. A typical daily cycle of color change was also observed. Injection of hemolymph from a dark animal into a light one resulted in darkening of the recipient. The reciprocal experiment produced negative results. Cauterization of the head also resulted in blanching of dark individuals whereas cauterization of the first thoracic segment resulted in pigment dispersion. He (Okay, 1945a) later found that extracts of the head and the first thoracic segment produced concentration of melanophore pigment. Boiling the extracts enhanced their pigment-concentrating activity. In still another study (Okay, 1945b) he found that extracts of the heads of *Idotea balthica, Ligia italica, Tylos latreilli,* and *Armadillidium granulatum* induced concentration of the pigment in melanophores of *Idotea balthica.* The heads of *Tylos* and *Armadillidium,* however, were not capable of causing concentration of pigment until they were boiled.

Suneson (1947) found that extracts of heads from *Idotea neglecta* and *Idotea emarginata* induced concentration of pigment in the melanophores of dark specimens. No dispersing effect was observed when these extracts were injected into pale specimens. Carstam and Suneson (1949) found that extracts of tails as well as heads of *Idotea neglecta* concentrated melanin in *Idotea* kept on a black background. Occasionally melanin dispersion was observed 60–120 min after injection of head and tail extracts into *Idotea* maintained on a white background. In addition, when head extracts were injected into specimens of *Palaemon squilla* on a white background the pigment in the red chromatophores dispersed but only the yellow pigment in the red-yellow chromatophores dispersed. Tail extracts dispersed all of the pigments. Furthermore, when extracts were injected into eyestalkless *Palaemon* the red pigment in the red-yellow chromatophores all over the body and the pigment in the red chromatophores of the carapace, but not of the abdomen, concentrated. Tail extracts had no effect on eyestalkless *Palaemon.*

Trachelipus rathkei, a terrestrial isopod, shows weak physiological color changes (McWhinnie and Sweeney, 1955). As is typical of isopods, the melanophores of *Trachelipus* show a primary response to light; the black pigment disperses further with increased illumination. Dispersion of the red pigment of the crayfish *Cambarus* sp. was induced by extracts of the sinus glands, optic lobes, and cerebral ganglia of *Trachelipus,* whereas extracts of the circumesophageal connectives or any segment of the thoracic nerve cord induced strong concentration of the same pigment (Fig. 19). Responses of *Trachelipus* to injected extracts were inconclusive but did suggest that the reactions of the pigmentary system of *Trachelipus* are opposite to those of *Cambarus*.

Oguro (1959c) observed a daily rhythm of pigment dispersion in the melanophores in the marine isopod *Idotea japonica*. The pigment was more dispersed by day than at night as in other isopods. This rhythm was apparent in specimens in darkness and in light on a black background but was completely inhibited in specimens on a white background under an illumination of

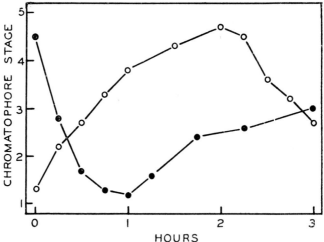

FIG. 19. Responses of red chromatophores of *Cambarus* to extracts of organs from *Trachelipus*. Circles, sinus glands, optic tracts, and cerebral ganglia. Dots, circumesophageal connectives and thoracic nerve cord. (Redrawn from McWhinnie and Sweeney, 1955.)

60 m.c. The amplitude of the cycle was greater in specimens in darkness than in light on a black background as was reported for the rhythm of *Ligia exotica* by Fingerman (1956b). Extracts of the ordinary sinus gland, accessory sinus gland, and cerebral ganglia dispersed the pigment in melanophores. On the other hand, extracts of the circumesophageal connectives and ventral nerve cord had no effect on the melanophores. One finds two contradictory statements in Oguro's publication. In one portion of the paper, page 250, Oguro (1959c) stated, "Some of the

extracts had scarcely any melanophore-concentrating action. Every injection brought about slight concentration of melanophores in a few minutes after the injection, but in a short time the melanophores fully expanded. These results are similar to those obtained by Enami (1941b) in *Ligia exotica*. He observed that injection of a head extract of *L. exotica* called forth a temporary concentration of melanophores." Then on page 251 Oguro stated, "In this animal (i.e. *Idotea japonica*) a melanophore-expanding factor only is found and no melanophore-concentrating one."

Armitage (1960) reported that the melanophores and xanthophores of *Ligia occidentalis* exhibit a daily cycle of pigment migration. In this species also, the pigment in both chromatophores was more dispersed by day than at night. Within one day an illumination of 150 ft c. suppressed the expression of the rhythm in specimens on black and on yellow backgrounds. The pigment in melanophores was more dispersed in specimens acclimated to a black background than to a yellow one. The yellow pigment, however, was less dispersed on a black background than on a yellow one.

The best possible explanation of the conflicting results obtained by different investigators of chromatophorotropins in isopods, is that the melanophores are regulated by pigment-dispersing and pigment-concentrating principles, and that the relative amount of such substance varies from organ to organ and from species to species. Consequently, in some instances the obvious effect of injected extracts is concentration of the black pigment and in others dispersion. The same situation may well be true of higher crustaceans also. The other chromatophores in isopods, xanthophores and guanophores, probably have their own dual control by different chromatophorotropins. Further experiments are needed to substantiate this hypothesis. The explanation of why some investigators have found that illumination inhibits the cycle of color change and others have not may be that a critical intensity of illumination exists above which the expression of this rhythm is inhibited.

CHROMATOPHORES OF NATANTIANS

Historically the natantians were the crustaceans whose chromatophores were first shown to be under hormonal rather than nervous control (Koller, 1925, 1927, 1928; Perkins, 1928). Representatives of three genera, *Crangon*, *Palaemonetes*, and *Palaemon*, have been studied in detail. The results of these studies will be emphasized below but not to the exclusion of the small amount that has been done on representatives of other genera of natantians.

The mechanism of regulation of the chromatophores in *Crangon* was described in a series of papers by Brown and his associates. Brown and Ederstrom (1940) observed that *Crangon* was usually light upon a white background and dark upon a black one. When both eyestalks were removed the bodies lightened and the tails darkened immediately after the operation. After about 30–45 min. these blinded individuals had an intermediate, mottled appearance, that would be maintained for several days. Injection of extracts of the circumesophageal connective ganglia into eyestalkless animals produced a striking, rapid darkening of the telson and uropods, followed by complete blanching. The ganglia of the circumesophageal connectives in *Palaemonetes* contained a substance that produced the same effect in *Crangon*, but no evidence was found for the existence of a comparable principle in the crabs *Uca*, *Carcinus*, *Pagurus*, or *Libinia*. A principle in the sinus gland of *Crangon* and *Palaemonetes* antagonized the action of the darkening material from the circumesophageal connectives. Such a principle was absent from the sinus glands of *Carcinus*, *Uca*, and *Pagurus*.

In 1940, Brown and Scudamore reported the results of a comparative survey of the effects of extracts of sinus glands from seven species of crustaceans (*Callinectes sapidus*, *Carcinus maenas*, *Crangon vulgaris*, *Libinia dubia*, *Pagurus pollicaris*, *Palaemonetes vulgaris*, and *Uca pugilator*) upon the red chromatophores of *Palaemonetes* and the black chromatophores of *Uca*. These extracts showed dissimilar relative effects upon the two types of chromatophores that cannot be explained in terms of different concentrations of a single chromatophorotropin. In support of this hypothesis, they showed that ethyl alcohol-soluble and insoluble fractions of the sinus glands differed in their relative chromatophorotropic effects upon the two chromatophore types. In view of the results of Brown and Scudamore that sinus glands have two chromatophorotropins and of Brown and Ederstrom that the sinus glands of *Crangon* have an active principle that is absent from the sinus glands of *Carcinus Uca, Pagurus,* and *Libinia,* two possibilities exist: (1) *Crangon* has three active substances in its sinus glands or (2) *Crangon* has two principles with one of the two different from either of those found in the other decapods that were investigated.

The circumesophageal connectives of *Crangon* possess two chromatophorotropins (Brown, and Wulff, 1941; Brown, 1946). One acts to lighten all the body except the telson and uropods, while the other darkens the telson and uropods and presumably in the absence of the former, the rest of the body as well. Brown and Saigh (1946) examined the central nervous organs of thirteen species of crustaceans representing the Isopoda, Natantia, Astacura, Anomura, and Brachyura for the presence of *Crangon* body-lightening hormone (CBLH) and *Crangon* darkening hormone (CDH). CBLH was found in each species. CDH was absent from the central nervous organs of brachyurans alone. Brown and Klotz (1947) were able to separate CDH and CBLH from one another by chemical means and thereby conclusively demonstrate that (1) these two principles exist and (2) the assumption that CDH in the absence of CBLH will darken the telson, uropods, and body is correct.

More recently, Sandeen and Fingerman (1959) studied the electrophoretic behavior of the antagonistic hormones controlling color changes in *Crangon septemspinosus.* Extracts of circumesophageal connectives with the tritocerebral commissure attached were submitted to electrophoresis at 500 V and 0.1–0.2 mA with citrate-phosphate buffer or borate buffer. At pH values of 5.2 and 7.8 both the tail-darkening hormone and the body-lightening hormone were electropositive. At pH 9.0 the greatest amount of tail-darkening hormone was recovered from the origin while the greatest amount of body-lightening hormone was recovered from the first inch toward the cathode. Tail-lightening hormone from eyestalks also migrated toward the cathode at these pH values.

With respect to *Palaemonetes*, the early studies of Perkins (1928) and Brown (1933, 1935a, b) showed that extracts of eyestalks and central nervous organs outside the eyestalk caused concentration of red pigment. In the meanwhile, Perkins and Kropp (1932) had shown that extracts of *Palaemonetes* eyestalks darkened frog tadpoles, *Rana clamitans*. Brown (1940) showed that the sinus gland of *Palaemonetes* accounts for approximately 80% of the red pigment-concentrating activity of the eyestalk.

Brown, Webb, and Sandeen (1952) found two antagonistically functioning chromatophorotropins in central nervous organs of *Palaemonetes*; one concentrated the red pigment and the other dispersed it. The duration of the effect of the concentrating hormone was shorter than that of the dispersing hormone. A given concentration of dispersing substance was more effective in dispersing pigment in one-eyed animals than in intact ones when both were maintained on a white background. This observation was interpreted in terms of a reduced amount of concentrating substance that could be secreted by one-eyed animals in response to the background. When tritocerebral commissures and circumesophageal connectives were tested separately, the latter were found to possess practically all of the dispersing factor, while both parts possessed considerable amounts of the concentrating material. The sinus gland contained little or none of the dispersing factor shown to be present in the entire eyestalk. A comparison of the red pigment-concentrating activity of a sinus gland with that of a tritocerebral commissure indicated that these two sources possessed approximately equal amounts of concentrating principle. The abdominal cord possessed the highest ratio of dispersing factor to concentrating factor of any of the other central nervous organs.

Fingerman, Sandeen, and Lowe (1959) were interested in the effects of long-term adaptation upon (1), the rate of red pigment migration and (2), the endocrine sources. The rates of red pigment concentration and dispersion decreased progressively in specimens maintained on black and on white backgrounds for 2 weeks. During the 2 weeks that the prawns were on the black and the white backgrounds the titer of red pigment-dispersing hormone in the circumesophageal connectives changed. The quantity of red pigment-dispersing hormone increased in the circumesophageal connectives of specimens on a white background whose red pigment was maximally concentrated and decreased in the circumesophageal connectives of specimens on a black background whose red pigment was maximally dispersed. The hormone not being used to maintain the appropriate degree of pigment dispersion was stored and that which was used decreased as compared with the titers in the circumesophageal connectives of specimens that had been on black and on white backgrounds 2 hr. These experiments demonstrated that (1), the materials in central nervous organs outside the eyestalk are normally involved in physiological color changes and (2), the color change system becomes sluggish during a period of long-term stabilization. Fingerman, Sandeen, and Lowe (1959) also considered other aspects of the control of *Palaemonetes* chromatophores. Boiled extracts of eyestalks, optic ganglia, supraesophageal ganglia with the circumesophageal connectives attached, and abdominal nerve cords produced more dispersion of red pigment than did unboiled extracts. One possible explanation of this finding is that the red pigment-dispersing hormone exists in two forms, in an inactive state in heat-labile neurosecretory granules and in the free active state. In addition, the electrophoretic behavior of these chromatophorotropins was determined at pH 7.2. The red pigment-concentrating hormone in the eyestalk was electropositive and the dispersing substance electronegative.

Sandeen and Costlow (1961) assayed the nervous systems of three barnacles (*Balanus eburneus*, *Chelonibia patula*, and *Lepas* sp.) on *Palaemonetes vulgaris* and found no red pigment-concentrating material but a small amount of a red pigment-dispersing substance. This seems to be the first report of the occurrence in an organism of a red pigment-dispersing substance for *Palaemonetes* without a red pigment-concentrating one. The function of this substance in barnacles, none of which have chromatophores, should be determined. Neurosecretory cells of unknown function had previously been observed in the cirripedes *Pollicipes polymerus*, *Balanus glandula*, *B. hesperius laevidomus*, *B. nubilis*, *B. rostratus*, and *Chthamalus dalli* by Barnes and Gonor (1958).

Very little research has been done on the control of chromatophores of larval crustaceans. Broch (1960) found that the red pigment in the first zoea of *Palaemonetes* was more concentrated in specimens on a white background than on a black one, just as occurs in the adult. Extracts of supraesophageal ganglia and eyestalks from adults caused concentration of red pigment in the first zoea; extracts of adult abdominal nerve cords produced dispersion of the red pigment. A cytological study of developing *Palaemonetes* would be interesting in order to learn at what larval stage functional neurosecretory cells are present and whether these cells contain chromatophorotropins. In contrast, Pautsch (1953) had found previously that chromatophores in the zoea of *Crangon crangon* were not responsive to chromatophorotropins from adults.

In 1946 Panouse published the results of his extensive investigation of the chromatophore system of *Palaemon serratus*. He reported that the sinus gland and central nervous organs each contained a substance that would concentrate red pigment. Panouse was also able to disperse the red pigment with extracts of the supraesophageal ganglia, but he thought the latter effect was a nonspecific response and not due to a pigment-dispersing hormone. In view of the results of Brown, Webb, and Sandeen (1952), described above, it would appear that Panouse was unduly cautious. He thought the dispersing effect was nonspecific because (1), the degree of dispersion was not as great as the concentrating effect and (2), the concentrating effect was completed sooner than the dispersing effect. Both statements apply equally well to the red pigment-dispersing hormone of *Palaemonetes*. Further-more, these are the same characteristics shown by the red pigment-dispersing hormone of the dwarf crayfish, *Cambarellus shufeldti*, reported by Fingerman (1957a). Panouse (1946) also stated that the blue pigment of *Palaemon serratus* at Roscoff, France, where he did his experiments, appeared in the chromatophoral branches when the red pigment concentrated and that the blue pigment appeared to be formed from the red pigment. This observation is the same as that of Brown (1934, 1935a, b) concerning the origin of the blue pigment in *Palaemonetes vulgaris*. In contrast to the observation of Panouse (1946), Scheer and Scheer (1954) reported that specimens of the same species, *Palaemon serratus*, from Naples, Italy, possessed red and blue pigments that were not in the same chromatophores. The blue chromatophores are supposed to be concentric with the red ones. The findings of Scheer and Scheer (1954) are unique and certainly deserve further study.

The color pattern of *Palaemon serratus* was described by Know-les (1953). It depends on chromatophores arranged in a pattern of stripes separated by less intensely colored interspaces. The stripes are roughly longitudinal on the cephalothorax and trans-verse on the abdomen. Scheer and Scheer (1954, 1955) directed their efforts toward understanding the color changes in *Palaemon serratus* to four types of pigments, red in the stripes, blue in the stripes, red in the interspaces, and blue in the interspaces. These four pigments underwent small, independent, but statistically significant cyclic changes in degree of dispersion during the inter-molt cycle. The mean length of the intermolt period of intact specimens was 7.6 days. The degree of dispersion of each pigment, as well as the period of time the pigment was in that state of dispersion, could be correlated with the duration of one or more stages of the intermolt cycle, thereby suggesting that the hormonal factors which control these four pigments are also concerned with the metabolic processes of the intermolt period. Extracts prepared from eyestalks and from the entire anterior portion of the central nervous system (supraesophageal ganglia, circum-esophageal connectives, and thoracic nerve cord) caused concentration of each of the four pigments, and, in addition, dispersed the blue pigment of the stripes. The anterior portion of the central nervous system had an additional effect, dispersion of the red pigment of the interspaces. Eyestalk removal resulted in increased dispersion of each pigment but did not completely abolish the cyclical color changes occurring during the intermolt period; presumably chromatophorotropins from the remaining central nervous organs were responsible for these changes. Consideration of the relations between color and the duration of the molt cycle stages, and of the effect of eyestalk removal on color and the molt cycle, led Scheer and Scheer (1954) to the conclusion that at least five chromatophorotropins are concerned with specific events in the intermolt cycle, and that none of these principles is identical with the molt-inhibiting hormone of other crustaceans.

Scheer and Scheer (1954) also performed some brief experiments on the chromatophores of *Lysmata seticaudata*. These investigators found no response when extracts of eyestalks or central nervous organs outside the eyestalk were injected into *Lysmata* or *Palaemon serratus*. This result was certainly surprising and unexpected. However, Burgers (1959), working at Naples where Scheer and Scheer had done their experiments, found that extracts of eyestalks from *Lysmata seticaudata*, *Palaemon serratus*, and *Palaemon elegans* caused a distinct dispersion of the pigment in the crab *Macropipus vernalis*. In view of this conflicting evidence, the color change system of *Lysmata* certainly needs further investigation.

The apparent difference between the location of the blue pigment in the specimens of *Palaemon serratus* from the Roscoff and Naples populations may have arisen as a result of geographical isolation. A similar problem on geographical diversity has arisen with respect to molting. Carlisle (1954) found that removal of the eyestalks from specimens of *Palaemon serratus* at Plymouth. England, lengthened the intermolt period, whereas Drach (1944) had clearly shown that the intermolt period of specimens of this species collected at Roscoff decreased after eyestalk ablation. With specimens from Naples, Scheer and Scheer (1954) noted that molting was less frequent in eyestalkless than in intact specimens. In attempting to clear up this conflict, Carlisle (1955) found that populations of *Palaemon serratus* from Plymouth, from Roscoff on the north coast of France, and from Concarneau on the Atlantic coast of France, show readily distinguishable, characteristic differences in color pattern. However, he does not feel that these populations are subspecies nor did he find signs of sterility in crossbreeding experiments with members of the three populations.

Knowles (1952) found that the sinus gland of *Palaemon squilla* is ineffective on the white pigment whereas extracts of the entire eyestalk concentrated this pigment. Permanent dispersion of the white pigment was apparent in eyestalkless specimens, but this pigment would concentrate and disperse when sinus glandless specimens were placed on black and on white backgrounds respectively. Knowles (1953) later reported that extracts of the post-commissure organ of *Palaemon serratus* caused concentration of white pigment in this species.

Carlisle, Dupont-Raabe, and Knowles (1955) reported in preliminary form the results of their studies of the electrophoretic behavior of chromatophorotropins in *Palaemon serratus*. A subsequent detailed publication on the same subject was authored by Knowles, Carlisle, and Dupont-Raabe (1955) who found a relatively immobile material in the sinus gland and post-commissure organs that they called the A-substance. It was electropositive at pH 7.8 and concentrated the pigment in the large and small red chromatophores. Another principle with low mobility, called the B-substance, was found only in the post-commissure organs. The latter was electronegative at pH 7.8, concentrated the pigment in the large red chromatophores, but dispersed the pigment in the small red chromatophores. When extracts of tissues containing the A-substance were allowed to stand a short time prior to electrophoresis at pH 7.5, two oppositely charged substances that had not been detected previously were found. These were termed alpha-substances and possibly represented disintegration products of the A-substance. Their function was to concentrate the pigment in small red chromatophores.

Knowles and Carlisle (1956) and Carlisle and Knowles (1959) attempted to identify the A- and B-substances with chromatophorotropins found in other organisms. Such correlations are interesting but because we have so little information concerning the number of chromatophorotropins involved in the color change process in any one species of crustacean we should proceed with caution. In experiments of this sort we must face the possibility that two different substances with the same or very similar mobilities and the same charge are present on the filter paper strip and might, therefore, be considered one substance. Such an occurrence may account for the opposite effect of the B-substance on the large and the small red chromatophores, whereas the A-substance has one effect on both of these chromatophores. The red pigment-concentrating hormone in the eyestalk of *Palaemonetes* is electropositive at pH 7.2 (Fingerman, Sandeen, and Lowe, 1959) as is the A-substance of *Palaemon*. The red pigment-dispersing hormone from the eyestalk of *Palaemonetes* is electronegative just as is the B-substance of *Palaemon* one of whose functions is dispersion of the pigment in the small red chromatophores.

Pasteur (1958) reported that ablation of the X-organ from each eye of *Palaemon serratus* resulted in maximal dispersion of the red and yellow pigments. Her description of the operation suggests that it was the sensory pore X-organ she removed. The results were the same as if the sinus gland had been ablated, thereby giving some physiological evidence of axonal transport of chromatophorotropins from the X-organ to the sinus gland.

Very little research has been done with other natantians. Nagano (1943) found that the red pigment of eyestalkless specimens of the shrimp *Paratya compressa* remained permanently dispersed. Concentration of this pigment was effected by eyestalk extracts which also darkened tadpoles of *Rana nigromaculata* and lightened the fish *Oryzias latipes*. In 1950 Nagano reported that the chromatophores of *Paratya* are insensitive to a variety of drugs among which are atropine, morphine, and insulin. Knowles (1953) found that the tritocerebral commissure region of *Penaeus brasiliensis* contained substances that concentrated red and white pigments.

Knowles (1956) set forth some criteria that he recommended all investigators use to decide whether a color change is due to the presence of a chromatophorotropin or is a nonspecific response: (1) some response should occur within 5 min. (2) the response should last at least 30 min. (3) the response should be one that can be found in normal specimens under normal conditions; and (4) the injected substance should not be toxic. The last three appear to be sound. However, the first criterion may, in many cases, be difficult to adhere to, witness the delayed response to dispersing hormone in *Palaemon* (Panouse, 1946) and *Palaemonetes* (Brown, Webb, and Sandeen, 1952) wherein the concentrating effect occurs first and appreciable pigment dispersion is suppressed until the concentrating effect starts to wane.

One point appears to be emerging from the rapidly increasing volume of data about control of chromatophores in Natantians; the dark chromatophores are regulated by pigment-dispersing and pigment-concentrating substances. Evidence exists for a white pigment-concentrating hormone, but not for a white pigment-dispersing principle. When the time arrives that we can purify and identify these chromatophorotropins chemically than we shall be able to learn how many different chromatophorotropins exist or whether, for example, all natantians produce but one chemical that is effective in concentrating red pigment or whether several different substances exist. An unsolved problem of great importance is whether the circulating form of the hormone is the same as the one that is extracted from neurohaemal organs.

CHROMATOPHORES OF BRACHYURANS

The general principle that the Brachyura (true crabs) become pale when their eyestalks are removed was set forth by Brown (1948a). This was true of all crabs that had been investigated up to that time. Later, however, Enami (1949, 1951b) reported an exception to this rule. He found that removal of both eyestalks from specimens of three species of *Sesarma*, namely *intermedia*, *haematocheir*, and *dehaani*, resulted in permanent dispersion of the dark pigments. Enami's work will be discussed in detail below, but first some of the research with crabs that blanched after eyestalk removal will be considered.

Carlson (1935, 1936), who seems to have been the first to investigate the endocrine regulation of chromatophores in crabs, noted that removal of the eyestalks from *Uca pugilator* and *U. pugnax* resulted in blanching, concentration of melanin. Injection of eyestalk extracts resulted in melanin dispersion, darkening. Sandeen (1950) studied in detail the endocrine control of the black and the white chromatophores of *U. pugilator*. She found two substances, one dispersed the black pigment and the other concentrated the white pigment. An antagonism between these two principles was apparent. A large amount of melanin-dispersing hormone decreased the expression of the white pigment-concentrating principle. Sandeen obtained no direct evidence for a melanin-concentrating principle.

Earlier, Enami (1943) had studied the pigmentary system of the fiddler crab, *U. dubia*, and reported that eyestalk removal resulted in blanching as occurs in *U. pugilator* and *U. pugnax*. Surprisingly, he also reported that the sinus gland and central nervous system contained a black pigment-concentrating hormone but obtained no evidence for a black pigment-dispersing substance. The results of Enami (1943) are subject to criticism because he did not have a control group in any of his experiments. This writer has observed (unpublished) that injection of saline into intact *U. pugilator* will produce as much pigment concentration as did the extracts used by Enami.

Brown (1950) found that eyestalk removal resulted in maximal concentration of red pigment in *U. pugilator*. The sinus glands and all portions of the central nervous system of *U. pugilator* contained red pigment-concentrating and red pigment-dispersing principles. The action of the red-concentrating principle could dominate the response when both substances were present in high concentrations, with the result that maximal concentration preceded maximal dispersion. The concentrating effect had a shorter duration than the dispersing effect. The behavior of these antagonists in mixtures is the same as was discovered later for the red-concentrating and red-dispersing substances in *Palaemonetes* by Brown, Webb, and Sandeen (1952) and in *Cambarellus shufeldti* by Fingerman (1957a). Brown and Fingerman (1951) showed by extraction of supraesophageal ganglia with absolute isopropyl alcohol that the red pigment-dispersing substance in *Uca pugilator* was not identical with the black pigment-dispersing substance. A possible identity of the principles had been suggested by Brown (1950). The alcohol-soluble fraction had much black-dispersing but little red-dispersing activity. The reverse was true of the alcohol-insoluble fraction.

Fingerman (1956c) considered the problem of the existence of a melanin-concentrating substance in fiddler crabs, because (1), the results of Enami (1943) with *U. dubia* are subject to criticism, and (2), Sandeen (1950) had found no evidence for such a substance in *U. pugilator*. Brown and Stephens (1951), Brown and Hines (1952), Webb, Bennett, and Brown (1954), and Hines (1954) had postulated a *Uca*-lightening hormone in *U. pugilator* and *U. pugnax*. The results of their experiments could be explained more simply by assuming that body-lightening was due to a melanin-concentrating hormone rather than to removal of the darkening hormone from the circulation. By perfusion of isolated legs with blood or sea water, Fingerman (1956c) was able to demonstrate the presence of (1), a black pigment-dispersing hormone in the blood of *U. pugilator* that had maximally dispersed melanin and (2), a black pigment-concentrating hormone in the blood of crabs with maximally concentrated melanin. Maximally dispersed melanin in the chromatophores of a leg of *Uca* will gradually concentrate when the leg is removed from the body (Hines, 1954). Fingerman (1956c) found that the rate of melanin concentration was slowed by perfusion of blood from dark specimens and increased by blood from specimens whose black pigment was maximally concentrated. Control legs were perfused with sea water. These results could not have been obtained unless a lightening factor were in the blood of the pale *Uca* and a darkening factor in the blood of the dark crabs. As is becoming evident with more and more investigation, the chromatophores of all crustaceans appear to be controlled by pigment-dispersing and pigment-concentrating substances.

Fingerman and Fitzpatrick (1956) showed that the pigment in melanophores of female *Uca pugilator* collected at Ocean Springs, Mississippi, was more dispersed than the melanin in males. Removal of the large cheliped from males resulted in approximately equal coloration of males and females. Furthermore, the greater the number of appendages removed from both sexes, the darker the crabs. Presumably, removal of the appendages resulted in a decreased circulatory space so that the melanin-dispersing hormone could not become as diluted as in intact crabs. These experiments in which legs were amputated were performed during a phase of the rhythm of color change (Abramowitz, 1937a) when one would expect the blood to contain much melanin-dispersing hormone and very little of the melanin-concentrating principle.

Stephens, Friedl, and Guttman (1956) attempted by use of filter paper electrophoresis to isolate and characterize the chromatophorotropic principles in the sinus gland of *Uca pugilator*. They were able to distinguish three distinct peaks of black-dispersing activity, one dubious area of black-concentrating activity, and at least one peak of dispersing activity and one peak of concentrating activity for the red and the yellow chromatophores. At least two peaks of white-concentrating activity were discernible. The possibility that three different black pigment-dispersing substances occur in the sinus gland is intriguing. It would be interesting to learn which, if any, is the circulating form of the hormone or whether all three are released from the sinus gland.

Bowman (1949) found that the sinus gland of the grapsoid shore crab *Hemigrapsus oregonensis* contained a higher concentration of a material that dispersed melanin than did the optic ganglia. The melanin of eyestalkless specimens became punctate as occurs in *Uca*. Bowman also stated the melanin-dispersing substance is "possibly in the brain and thoracic mass of ganglia." The fact that Bowman did not show this conclusively is surprising, since the supraesophageal ganglia and thoracic ganglia of all other brachyurans which have been investigated contain chromatophorotropins. The probable explanation for this deficiency is the small quantity of tissue from the supraesophageal ganglia and thoracic ganglia that he assayed. The volume was equal to one-half a sinus gland of *Hemigrapsus*. This minute amount of nervous tissue obviously could not evoke a strong chromatophorotropic response, since the hormones are much less concentrated in the central nervous organs he used than in the sinus gland.

Matsumoto (1954b) studied the endocrine control of the chromatophores of the fresh water crab *Eriocheir japonicus*. Melanin of eyestalkless crabs was maximally concentrated, but the red pigment was maximally dispersed, the reverse of the condition of red pigment in eyestalkless fiddler crabs. The supraesophageal ganglia, thoracic ganglia, and eyestalks of *Eriocheir* contained a material that dispersed black pigment. Extracts of the thoracic ganglia and eyestalks concentrated red pigment.

The physiology of the black and the red pigment cells of the blue crab, *Callinectes sapidus*, was investigated by Fingerman (1956a). His results are similar to those of Matsumoto (1954b), but Fingerman analyzed some aspects of the system of *Callinectes* in greater detail than Matsumoto did with *Eriocheir*. Matsumoto, on the other hand, included an excellent study of the distribution and types of neurosecretory cells as described above, a subject Fingerman did not consider in *Callinectes*. As in *Eriocheir*, eyestalk removal resulted in maximal dispersion of red pigment. With regard to *Callinectes*, the sinus glands and central nervous organs contained material that dispersed black pigment and concentrated red pigment. Comparison of the order of decreasing potency of the organs that had been assayed on both black and red chromatophores with the activities of similar tissues of crabs reported in the literature showed no clearly defined trends. Determination of the alcohol solubility of the chromatophorotropins in the sinus glands and circumesophageal connectives of *Callinectes* revealed that the black pigment-dispersing hormone from the two sources had to be different in some way, whereas the red pigment-concentrating hormone from the two sources had the same alcohol solubility. The black pigment-dispersing hormone of the circumesophageal connectives was much less soluble in alcohol than the substance in the sinus gland that had the same function.

Burgers (1958) found that melanophores of *Uca rapax* behaved as the melanophores of *U. pugnax* and *U. pugilator*. Eyestalk removal resulted in maximal concentration of melanin. Melanin dispersion was produced by eyestalk extracts. The vertebrate hormones intermedin, adrenocorticotropic hormone, and epinephrine had no effect on the melanophores of *U. rapax*. In 1959 Burgers reported the results of studies of the erythrophores and melanophores of the crab *Macropipus vernalis*. Eyestalk ablation resulted in melanin concentration but did not affect the erythrophores. Injection of eyestalk extract resulted in dispersion of melanin but had no effect on the red pigment of intact or eyestalkless specimens. The red pigment appeared to be sensitive only to direct illumination. This pigment became more concentrated with increased illumination. The melanophores and erythrophores Burgers dealt with lie so close to one another that they form a color unit.

Enami (1951b) described the control of the chromatophores of three species of *Sesarma*, namely *intermedia*, *haematocheir*, and *dehaani*. He observed that after eyestalks were removed the black pigment became maximally dispersed and the white nearly maximally concentrated. The red and vermilion pigments in eyestalkless crabs attained an intermediate condition. The behavior of these pigments is obviously different from what has been observed in the other genera of crabs that have been investigated. Extracts of the sinus glands and central nervous organs from juveniles concentrated the red and vermilion pigments. Extracts of the medulla terminalis and other central nervous organs of juveniles produced concentration of black pigment and dispersion of white pigment. Sinus gland extracts were ineffective on the melanophores.

Fingerman, Nagabhushanam, and Philpott (1960c, 1961) decided to investigate the control of the melanophores in an American species of *Sesarma*, namely *reticulatum*, in order to make a comparison with the results of Enami (1951b). Eyestalkless specimens were exposed for 2 hr to a series of incident light intensities ranging from 2 to 1110 ft c. The degree of melanin dispersion in these eyestalkless crabs was a function of the incident light intensity (Fig. 20). The melanin was maximally concentrated at 2 and 20 ft c. With increased light intensity the

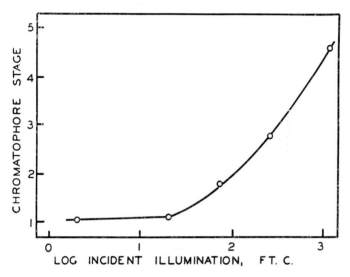

FIG. 20. Relationship between melanophore index of eyestalkless specimens of *Sesarma reticulatum* on a white background and the logarithm of the incident light intensity. (From Fingerman, Nagabhushanam, and Philpott, 1961.)

melanin began to disperse. This pigment was almost maximally dispersed in specimens exposed to an illumination of 1110 ft c. To determine the role of endocrines in pigment migration in *Sesarma reticulatum*, extracts were prepared of the sinus gland, optic ganglia, supraesophageal ganglia, circumesophageal connectives, and thoracic ganglia. These extracts were injected into eyestalkless *Sesarma reticulatum* whose melanin was maximally concentrated, and into specimens maintained under a high intensity of illumination so that the melanin was in a state intermediate between the fully dispersed and fully concentrated conditions. Each extract dispersed melanin. In Fig. 21 are shown the results

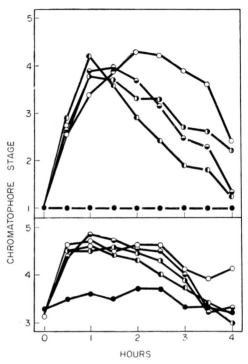

FIG. 21. Relationship between melanophore stages and time following injection of extracts of central nervous organs from *Sesarma reticulatum* into eyestalkless specimens in white containers under an illumination of 2 ft c. (upper portion of figure) and 560 ft c. (lower portion of figure). Optic ganglia, circles; supraesophageal ganglia, circles half-filled on bottom; circumesophageal connectives, circles half-filled on left; thoracic ganglia, circles half-filled on right; control, dots. (From Fingerman, Nagabhushanam, and Philpott, 1961.)

obtained with the central nervous organs. Extracts of sinus glands produced less dispersion than caused by any of the central nervous organs. In the case of the sinus glands, after the dispersing action of the extract, a small amount of pigment concentration was noticed in specimens whose pigment had been in an intermediate state at the time of injection of the extracts. Further comparisons might have been possible had Enami stated the intensity of illumination to which his eyestalkless animals were exposed. Another difference between the three species of *Sesarma* used by Enami and *Sesarma reticulatum* is that individuals of the latter species will exhibit color changes in response to black and to white backgrounds, especially at low light intensities, whereas Enami (1951b) stated that his crabs did not exhibit responses to background.

Brown and Cunningham (1941) showed that extracts of central nervous organs from the horseshoe crab, *Limulus polyphemus*, produced dispersion of black pigment and concentration of white pigment in *Uca pugnax*. The distribution of active material correlated with the distribution of neurosecretory cells in the central nervous system of *Limulus* as determined by Scharrer (1941). Sandeen and Costlow (1961), in addition to their experiments on *Palaemonetes* that were mentioned above, assayed barnacle nervous systems on *Uca pugilator*. A melanin-dispersing substance was found that was heat-stable and inactivated by trypsin.

Rhythms of color change have been described in more detail for crabs than for any other group of animals. These cycles are presumably due to rhythmical release of chromatophorotropins from the sinus glands and central nervous organs and for this reason will be considered in detail herein. Aspects of this subject have been considered in several recent reviews (Stephens, 1957a; Fingerman, 1957b, 1959a, 1960; Brown, 1957, 1958; Webb and Brown, 1959). Abramowitz (1937a) first described the 24-hr rhythm of color change in *Uca*. The crab was dark by day and pale by night. Brown and Webb (1948) found that the rhythm persisted for as long as 30 days in constant darkness and that the frequency was unaffected by temperatures from 6° to 26°C, but the amplitude decreased with decrease in temperature. When the rate of metabolic processes was greatly reduced by temperatures at 0°–3°C, the 24 hr rhythm of *Uca* was delayed by an interval

5

closely approximating the time of exposure to the low temperature. The same investigators (Brown and Webb, 1949) analyzed
further the 24 hr rhythm of color change in *Uca* and found that (1),
the phases of the chromatophore rhythm could be reversed by
illuminating animals at night and keeping them dark by day, (2),
exposure to 6 hr periods of illumination alternating with 6 hr
of darkness resulted in a 24 hr rhythm that was 6 hr out of
phase with solar day-night, and (3), such a rhythm, 6 hr
out of phase, may persist for several days in constant darkness and then gradually return to the previously established
rhythm if the last period of illumination occurred when the animals were entering the night phase of the rhythm. The shifted
rhythm, however, showed no persistence when the last period of
illumination occurred when the animals were entering the day
phase of the rhythm. Further analysis of the results led to the
adoption of an hypothesis involving two centers of rhythmicity
in the fiddler crab, each one capable of having its rhythm altered
independently of the other, and with one of the centers influencing
the second, which is in turn responsible for the rhythmical release
of chromatophorotropins. Webb (1950) showed that the basic
24 hr frequency of the rhythm of color change in *Uca* could not
be changed by exposure to alternating light and dark periods of
16 hr each. She also clarified further the mechanism underlying
phase shifts by light and dark changes. When animals showing
a normal rhythm were exposed to illumination beginning at
1.00 a. m. and ending either 6 or 12 hr later, a shift in the phase
of the rhythm occurred such that the most concentrated condition of the melanin occurred about 6 hr earlier than normal. This
phase shift persisted as long as the animals were kept in constant
darkness. When crabs were exposed to a 16, 20, or 24 hr period
of illumination which ended at 7.00 a.m., a persistent shift in
the phase of the rhythm occurred. The shift was such that the
most concentrated condition of the pigment occurred about 6 hr
later than normal. Brown and Stephens (1951) found that changes
in length of photoperiod induced persistent alterations in the
amplitude of the 24 hr cycle of color change in *Uca*. The greater

the photoperiod, the greater was the amplitude of the rhythm. These investigators also postulated two centers of rhythmicity controlling the melanophores, quite similar to the hypothesis of Brown and Webb (1949), but Brown and Stephens (1951) went further and postulated that one of the two centers would respond to darkness by calling for secretion of a black pigment-concentrating principle and to light by evoking release of a black pigment-dispersing hormone.

Brown and Hines (1952) showed that the amplitude of the 24 hr rhythm in *Uca* exposed to constant illumination varied inversely with the intensity of illumination. The reduction in amplitude was primarily the result of a decrease in the amount of melanin concentration that occurred during the night phase of the rhythm. Brown, Fingerman, and Hines (1954) studied further the mechanism involved in shifting the phases of the 24 hr rhythm in *Uca* by subjecting crabs to a series of combinations of brighter illumination by night and dimmer illumination by day. A graded series in the amount of shift of the phase of the rhythm with respect to solar day-night was obtained that was capable of being interpreted in terms of two operating factors: (1), the strength of the stimulus in the form of the increase in light intensity at 7.00 p.m. and (2), the absolute brightness of the higher illumination during the period from 7.00 p.m. to 7.00 a.m.

Webb, Bennett, and Brown (1954) found a 24 hr rhythm of chromatophorotropic response in eyestalkless *Uca pugilator* which they suggested might be due to cyclically varying quantities of black pigment-concentrating hormone in the blood. Extracts of eyestalks were less effective in dispersing black pigment of eyestalkless specimens at night than during the daytime, presumably because of rhythmical release of more melanin-concentrating hormone at night than during the day. During the night-time, melanin in intact specimens normally concentrates.

Brown, Bennett, and Ralph (1955) presented evidence for a reversible influence of cosmic ray showers on the chromatophore system of *U. pugnax*. Increased concentrations of cosmic ray showers resulted in increased pigment concentration during the initiation of transition into the day phase of the 24 hr cycle and increased melanin dispersion during at least most of the remaining hours of the day. No evidence has been presented as yet that the capacity to exhibit a response to alterations in intensity of cosmic ray showers is in any way normally operative in the maintenance of the precise 24 hr cycles of color change. These investigators decided to study the effects of cosmic radiation on chromatophores because it is known to be rhythmic (Forró, 1937; Barnathy and Forro, 1939).

Webb, Brown, Bennett, Shriner, and Brown (1956) demonstrated that population size is important for the expression of the rhythm of color change in *Uca*. When individuals were isolated and exposed to a constant low intensity of illumination the amplitude of the rhythm decreased markedly. When these animals were recongregated the amplitude of the rhythm rapidly returned to normal. Stephens (1957a) confirmed this observation and, in addition, demonstrated that the rhythm would return to normal in an intact crab if it had the company of even an eyestalkless crab whose melanin remained maximally concentrated. In another publication, Stephens (1957b) reported the results of his investigation of phase shifts by exposure of animals to sudden changes of temperature. He found that the 24 hr melanophore cycle of *U. pugnax* could be shifted out of phase with solar day-night cycles by exposing animals maintained in darkness to temperatures between 9.5° and 18°C during the summer when the water temperature is about 28°C. The amount of the shift appeared to depend on the time of day the crabs were first exposed to the lower temperature and also on the time of day they were warmed. To produce a shift, the low temperature had to be maintained for at least a minimum time, longer than 6 hr but not longer than 12 hr. Summation of shifts was also observed.

Twenty-four hour cycles of color change have also been observed in *Callinectes sapidus* by Fingerman (1955, 1956a) and in *Macropipus vernalis* by Burgers (1959). In both crabs the melanin was more dispersed by day than at night. In *Calli-*

nectes the red pigment showed a similar cycle. However, the red pigment of *Macropipus* did not show a daily rhythm under constant conditions.

Persistent tidal and semilunar rhythms of color change have been described in some crabs. Tidal rhythms have a frequency of 12.4 hr and semilunar cycles a frequency of 14.8 days. The first report of such rhythms of color change was published by Brown, Fingerman, Sandeen, and Webb (1953) for the fiddler crab, *Uca pugnax*. These crabs darkened by day and lightened by night in accordance with their 24 hr rhythm of color change. The tidal rhythm was evidenced by supplementary dispersion of the melanin 1–3 hr prior to the time of low tide. When low tide occurred early in the morning the curve skewed to the left and when low tide occurred in the afternoon the curve skewed to the right (Fig. 22). The curve was symmetrical about noon when low tide occurred about 2 p.m. and was bimodal when the low tide in the morning and the low tide in the evening, 12.4 hr later, could both produce supplementary dispersion of the melanin that would be superimposed upon the pigmentary dispersion due to the 24 hr cycle. As a result of possessing rhythms with both 12.4 and 24.0 hr frequencies, the crabs also possess a 14.8 day cycle, the mean interval between days on which these two rhythms repeat similar time relations to one another.

To learn whether the phases of the tidal rhythm are set by the tidal changes on the beach or determined directly by the phases of the moon and only secondarily correlated with the tides on the beach, the tidal rhythms of fiddler crabs collected in regions with different tidal times were compared. Chapoquoit on Cape Cod and Lagoon Pond on Martha's Vineyard were selected as collection sites because low tide occurred about 4 hr later at Lagoon Pond than at Chapoquoit. In other words, low tide at any given hour of the day would occur at Chapoquoit about 5 days later than at Lagoon Pond, the number of days required for the tidal cycle to progress about 4 hours over the 24 hr cycle. The phases of the tidal rhythm of *Uca* from Lagoon

Pond averaged 4.9 days earlier than the phases of *Uca* from Chapoquoit. On the basis of a rate of 48.8 min per day, 4.9 days is equivalent to 4 hr, the tidal difference between Lagoon

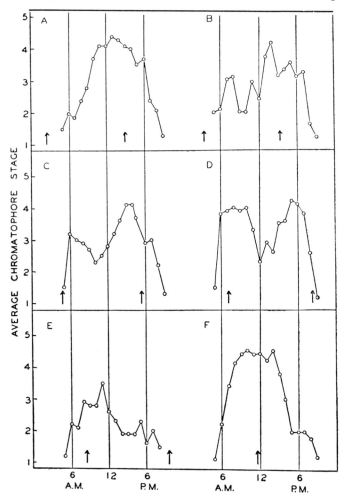

FIG. 22. Daily change in melanin dispersion of *Uca pugnax* in darkness on the day following collection. Arrows, times of low tide on days observations of the melanophores were made. A, Aug. 6; B, July 24; C, Aug. 9; D, Aug. 12; E, July 31; F, Aug. 3; 1952. (From Brown, Fingerman, Sandeen, and Webb, 1953.)

Pond and Chapoquoit. Evidently the phases of the tidal rhythm are determined directly by the local tidal situation.

An experiment was also performed to learn whether the 24 hr rhythm can be shifted in phase without altering the phases of the tidal rhythm. To solve this problem, *Uca* were collected and divided into two lots. One lot, the control, was placed in constant darkness. The phases of the 24 hr rhythm of the second group were shifted abruptly backward by three consecutive 12 p.m.–6 a.m. periods of illumination. Analysis of the 24 hr and tidal rhythms of both groups revealed that the 24 hr rhythm had been shifted backward 4.9 hr and the tidal rhythm 4.6 hr. The tidal rhythm, therefore, appears to be functionally associated with the 24 hr rhythm, because shifting the latter produces corresponding phase shifts in the former.

The amount of pigment concentration occurring in the melanophores of isolated legs of *Uca pugnax* depends upon the time of day that the legs are removed (Hines, 1954). During the day phase of the 24 hr rhythm when the melanin is dispersed less concentration occurs in legs autotomized near the time of low tide than near the time of high tide. Brown, Webb, Bennett, and Sandeen (1954) showed that the tidal rhythm of color change of *Uca pugnax* is temperature-independent between 13° and 30°C.

Evidence for an endogenous component of the daily and tidal rhythms of color change of *Uca* was presented by Brown, Webb, and Bennett (1955). The cycles of *Uca* shipped from Woods Hole, Massachusetts, to Berkeley, California, within a 24 hr period showed no tendency to drift away from the controls maintained in Woods Hole. The crabs were able to mark off quite accurately periods of solar and lunar day-lengths.

The persistent tidal and semilunar rhythms of color change in the blue crab, *Callinectes sapidus*, (Fingerman, 1955) are similar to those of *Uca pugnax*. The *Callinectes*, however, were collected in a region of daily tides (Lake Pontchartrain, Louisiana), whereas the *Uca pugnax* were from the Woods Hole Area where the tides are semidaily. The tidal rhythm of *Callinectes* had a 12.4 hr frequency, just as the tidal rhythm of *Uca pugnax*. The time between successive low tides in a region with diurnal tides is 24.8 hr, however. Evidently the center of tidal rhythmicity in *Callinectes* operates solely on the basis of tides spaced 12.4 hr apart, independent of the nature of the tides, high or low. The tidal rhythm of *Uca pugnax* is set to exert its maximal effect on the melanophores near the times of low tide whereas the blue

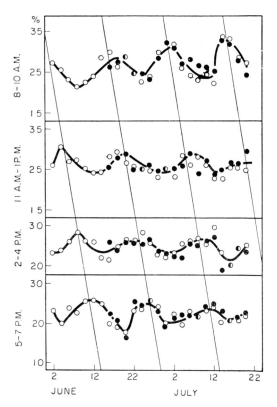

FIG. 23. Relationship between the percentage of the daily melanin dispersion of *Uca pugilator* occurring at each of four periods during the day and the day of the month. Note the tidal maxima passing over the daily periods at the tidal rate of approximately 50 min per day. There is also a 14.8-day cycle, the interval between each of the parallel diagonal lines. Circles, June 1, 1955, collection; dots, June 15, 1955, collection. (From Fingerman, 1956d.)

crabs showed no difference between their rhythmical responses at times of high and low tides.

Tidal and semilunar rhythms of color change were observed in another species of fiddler crab, *Uca pugilator*, by Fingerman (1956d). The specimens were collected at Ocean Springs, Mississippi, another beach with daily tides. The characteristics of both cycles were essentially the same as in *Uca pugnax* and *Callinectes sapidus* (Fig. 23).

Fingerman (1957c) and Fingerman, Lowe, and Mobberly (1958) continued analyzing the tidal rhythm of color change in *Uca pugilator* with emphasis on the role of the environment in setting the phases of the tidal rhythm relative to the 24 hr cycle of color change. Phase differences were observed between groups of crabs from burrows at different levels of the intertidal zone that correlated with the time required for the receding water to pass from one set of burrows to a set closer to the low tide mark. The hypothesis formulated was that the phases are set primarily according to the time the area was uncovered by the receding water and secondarily according to the time required for the area to drain so that the substrate would be firm enough to support openings to the surface from the burrows. The tidal rhythm would impart information to the crabs concerning the time when the portion of the beach above their burrows is in a condition favorable for them to forage for food.

CHROMATOPHORES OF ASTACURANS

Until recently, very little information was available concerning the chromatophores of Astacurans. In 1938 Kalmus reported that the white pigment of the crayfish *Astacus astacus* displayed a daily rhythm of migration. In intact specimens by day this pigment was more dispersed than at night. The rhythm persisted in eyestalkless individuals, but was 180° out of phase with the rhythm of intact specimens. Presumably, the rhythm was due to periodic release of chromatophorotropin from organs outside the eyestalk.

Brown and Meglitsch (1940) showed that the sinus gland of the crayfish *Orconectes immunis* contains two chromatophorotropins, white pigment-dispersing and red pigment-concentrating. An interesting assay technique utilizing chromatophores on isolated pieces of the carapace was designed by these investigators. A great advantage of this method is that the investigator does not have to contend with the hormones already present in the blood. McVay (1942), also working with *Orconectes immunis*, demonstrated that the central nervous organs contain appreciable quantities of red pigment and white pigment-concentrating principles. When both eyestalks were removed, both pigments dispersed maximally. She was able to show that two different substances were involved. The white substance in fresh, undried tissues was almost completely adsorbed on filter paper and entirely adsorbed on charcoal, whereas the red factor was not adsorbed on filter paper and only partially on charcoal. The white factor in dried tissues was not adsorbed upon filter paper. It was postulated that in fresh tissues the red and white active factors exist largely in combination with inactive carriers. The white combination would be adsorbed upon filter paper whereas the active group alone is not. Both forms are adsorbed upon charcoal. The red combination would be adsorbed upon charcoal whereas the active group alone is not. In dried tissues these combinations would have been broken down completely. It is worthwhile to compare these early results with the results of Pérez-González (1957) who presented evidence for release of hormone from neurosecretory granules. The combination postulated by McVay (1942) can perhaps be analogized with the neurosecretory granule plus its contained chromatophorotropin.

Bowman (1942) reported the results of an interesting study of morphological color change in *Procambarus clarki*. He counted the number of red and white chromatophores in the telson of young specimens at the time they were placed on a white or a black background and 52 to 110 days, later. The average number of red chromatophores in the telson of crayfish that had been on white was about 4 times the number of white ones whereas in specimens on black there were about 10 times as many reds as whites. Apparently, dispersion of pigment leads to an increased number of chromatophores and concentration of pigment leads to a decrease in the number of chromatophores. In this instance in specimens on a black background, the red pigment would be dispersed and the white concentrated. The reverse would occur in crayfish on a white background. Furthermore, the average number of white chromatophores in relation to the total number of chromatophores decreased in specimens on the black background and increased in specimens on the white background.

The eyestalks, supraesophageal ganglia, circumesophageal connectives, thoracic nerve cord, and abdominal nerve cord of the dwarf crayfish, *Cambarellus shufeldti*, contain chromatophorotropins that disperse red pigment, concentrate red pigment, disperse white pigment, and concentrate white pigment (Fingerman, 1957a). The dispersing and concentrating hormones of both pigments can be shown to be present in the same extract by injecting it into two groups of crayfish, one with the pigment under consideration maximally dispersed and the other with this pigment maximally concentrated. The red and white pigments in this crayfish disperse maximally when both eyestalks are removed, just as in *Orconectes immunis*. The major effect of injection of eyestalk extract of dwarf crayfish on its red pigment is not concentration but dispersion in spite of the fact that this pigment becomes maximally dispersed when both eyestalks are removed. The eyestalks contain only a very small amount of red pigment-concentrating hormone relative to the amount of dispersing hormone. The circumesophageal connectives contain more red pigment-concentrating hormone than any other organ of the central nervous system, which was also true in *Palaemonetes* (Brown, Webb, and Sandeen, 1952). The maximal response to red pigment-dispersing hormone from the eyestalk of *Cambarellus* occurred sooner than the maximal response to the substance with the same action from the other structures which were assayed.

In order to learn something of the "resting states" of the red and white pigments, portions of the carapace of *Cambarellus* with the associated chromatophores were removed from crayfish and immersed in physiological saline, thereby removing the chromatophores from the hormonal influence of the blood. The red pigment then concentrated and the white pigment dispersed (Fingerman, 1957d). Evidently, the "resting states" for these pigments are opposite each other. Chromatophores in the lateral portions of the carapace were slower to arrive at the appropriate "resting state" than those in the dorsal position. In line with this observation, Brown and Meglitsch (1940) noticed that the chromatophores in the dorsal portion of the carapace of *Orconectes immunis* were more responsive to chromatophorotropins than chromatophores in the lateral portions. These investigators also found that the red pigment of specimens from a black background would remain dispersed after isolation. It would appear the "resting states" of the red chromatophores of the two crayfishes are opposite to one another. *Cambarellus* and *Orconectes* also differ significantly in another aspect, the eyestalks of *Cambarellus* contain much red pigment-dispersing hormone and very little concentrating substance, whereas the predominant chromatophorotropin in eyestalks of *Orconectes immunis* is the red pigment-concentrating one. Curiously, however, in spite of this difference, the red pigment of both crayfishes disperses maximally after eyestalk ablation.

Reciprocal blood transfusions between dwarf crayfish that had been on a black or a white background for 2 hr led to the hypothesis that the blood at all times contained substances that disperse and concentrate the red and white pigments and that the degree of pigment dispersion at any time was due to the relative quantity of each principle in the blood (Fingerman, 1957d; Fingerman and Lowe, 1957a). These results differed from those of Koller (1925, 1927) who observed that blood from a black shrimp, *Crangon vulgaris*, darkened a light animal, whereas blood from a white donor had neither a lightening nor a darkening effect when injected into either light or dark individuals.

The effects of long-term background adaptation on the chromatophore system of *Cambarellus shufeldti* were determined by Fingerman and Lowe (1957a), who showed for the first time that the ability to change color in a invertebrate is facilitated through active use of the chromatophore system and becomes sluggish after a period of disuse. These investigators also demonstrated the changes that occur in the endocrine sources, blood titers, and target organs during long-term stabilization of the chromatophores on a specific background. Specimens of *Cambarellus* were collected and maintained on black and on white backgrounds for 3 weeks. During this time the rates of red and white pigment concentration and dispersion after appropriate background changes progressively diminished. Likewise, the inherent tendencies of red pigment on isolated pieces of carapace to concentrate and of white pigment to disperse gradually disappeared (Fig. 24). In addition to this physiological

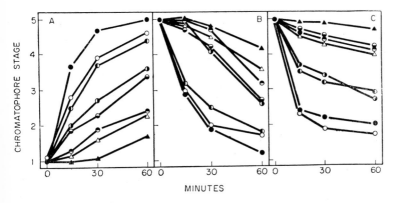

FIG. 24. A, responses of red chromatophores of specimens fo *Cambarellus* maintained for different periods of time on a white background and changed to black; B, responses of red chromatophores of specimens maintained on a black background and changed to white; C, changes in dispersion of red pigment in chromatophores isolated from the carapace of specimens kept on a black background. Dots, 2 hr; circles, 1 day; circles half-filled on left, 4 days; circles half-filled on right, 7 days; circles half-filled on bottom, 11 days; circles half-filled on top, 14 days; empty triangles, 18 days; solid triangles, 21 days. (From Fingerman and Lowe, 1957a.)

change, the red chromatophores of the crayfish kept on a black background changed morphologically; the number of processes in the chromatophores increased, the central body disappeared and the processes of adjoining chromatophores intermingled to such an extent that the chromatophores appeared to have lost their individuality when observed under the microscope.

To understand better the progressive decrease in ability to change color, the quantities of chromatophorotropins in the circumesophageal connectives and in the blood of specimens kept on black and on white backgrounds for 2 hr and for

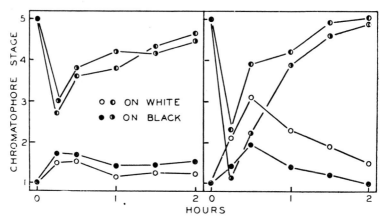

FIG. 25. Responses of red chromatophores of specimens of *Cambarellus* on black and on white backgrounds to extracts of circumesophageal connectives from specimens 2 hr on background (left) and 2 weeks on background (right). (From Fingerman and Lowe, 1957a.)

2 weeks were compared. The results showed that the quantity of the hormone not needed to maintain the appropriate degree of red pigment dispersion or concentration increased in the circumesophageal connectives (Fig. 25). For example, the amount of red pigment-dispersing hormone increased at least ten-fold in the circumesophageal connectives of specimens with maximally concentrated red pigment as a result of having been kept on a white background for 2 weeks. The titers of red pigment

activators in the blood also changed during the 2 weeks the crayfish were maintained on the black and the white backgrounds. Less of the hormone whose concentration in the circumesophageal connectives had increased was found in the blood, whereas the red pigment activator necessary to maintain the appropriate degree of background adaptation had increased.

Kleinholz (1957) had stated that all results obtained in studies of background adaptation in crustaceans could be explained with the chromatophorotropins in the eyestalk without invoking the secretory products of the supraesophageal ganglia or the circumesophageal connectives. The experiments described above provided the first evidence that the secretory products of central nervous organs outside the eyestalk play a role in the normal physiology of color change.

Fingerman and Lowe (1957b) were able to follow changes in the blood titer of red pigment-dispersing hormone after dwarf crayfish had been taken from a white background and put on a black one and vice versa. As expected, the titer increased upon transfer to black and decreased after transfer to white. These investigators initiated studies of the electrophoretic behavior of chromatophorotropins in *Cambarellus* with the ultimate objective of purifying and chemically characterizing these principles. At pH 7.4–7.8 the red pigment-dispersing principle in the supraesophageal ganglia and circumesophageal connectives was electronegative whereas the substance which concentrated this pigment was oppositely charged.

The rates of disappearance of chromatophorotropins from extracts of the eyestalks and the circumesophageal connectives of dwarf crayfish on standing at room temperature also were determined by Fingerman and Lowe (1957b). Red and white pigment-concentrating substances of the circumesophageal connectives disappeared at a much faster rate than the red and white pigment-dispersing hormones. The latter are presumably more stable substances. The red pigment-dispersing and white pigment-concentrating hormones were the first to disappear from eyestalk extracts.

Fingerman and Lowe (1957b) have tried to explain why in some crustaceans the dark pigment concentrates maximally and in others it disperses maximally after eyestalk removal. The ultimate condition of the pigment in chromatophores of crustaceans after eyestalk removal has been thought for a long time to depend on the chromatophorotropins produced by the eyestalks. For example, after removal of the eyestalks of the fiddler crab, *Uca pugilator*, black pigment becomes maximally concentrated. This condition was thought to be due to removal of the source of a black pigment-dispersing hormone. Results obtained with *Cambarellus* do not support this concept, however. Red pigment of eyestalkless *Cambarellus* is maximally dispersed, yet the predominant chromatophorotropin produced by the eyestalks that affects red pigment has a dispersing action (Fingerman, 1957a). The ultimate stage of the chromatophores of eyestalkless individuals is probably determined by hormones released from central nervous organs remaining after eyestalk removal and not primarily to the absence of chromatophorotropins from the eyestalk. The hormones that are not predominant quantitatively in the remaining central nervous organs, but are relatively stable molecules, probably determine the final stage of the chromatophores of eyestalkless crayfish.

Assays of mixtures of pigment-dispersing and pigment-concentrating principles which had been separated by electrophoresis revealed that the response depended upon both (1), the concentration of each chromatophorotropin and (2), the ratio of the concentrations of the antagonists. Furthermore, the character of the responses to mixtures depended upon the source of the chromatophorotropins. For example, the maximal response to dispersing-hormone from the eyestalk usually occurs sooner than the maximal response to dispersing-hormone from the supraesophageal ganglia and circumesophageal connectives. The degree of antagonism of red pigment-concentrating hormone which produces its maximal response at about the same time as the maximal response of dispersing hormone from the eyestalks was greater when mixed with the latter substance than with dispersing hormone from the supraesophag:al ganglia and circumesophageal connectives. As mentioned earlier, Fingerman and Mobberly (1960a) found that the isoelectric point of eyestalk red pigment-dispersing hormone is near pH 8·9 and that at pH 2·3 the red pigment-dispersing hormone in the supraesophageal ganglia and circumesophageal connectives is electropositive but electronegative at pH 7.7.

Knowles, Carlisle, and Dupont-Raabe (1955) found a red pigment-concentrating principle in the sinus gland and post-commissure organs of *Palaemon* that was electropositive at pH 7.8 and in addition another pigment-concentrating hormone in the post-commissure organs which was electronegative at pH 7.8. These findings highlight a difference between *Cambarellus* and *Palaemon*. In *Cambarellus* only one pigment-concentrating substance was found outside the eyestalk and the pigment-concentrating substances in the sinus glands of *Cambarellus* and *Palaemon* were oppositely charged.

The chromatophore system of another crayfish has also been investigated in detail. The red pigment of immature and adult *Orconectes clypeatus* dispersed maximally in specimens on a black background, but when the crayfish were transferred to a white background this pigment concentrated only to an intermediate degree (Fingerman. 1958). The white pigment, however, concentrated maximally and dispersed maximally in response to black and to white backgrounds respectively. Substances which caused dispersion and concentration of the red and white pigments were found. When chromatophores were removed from animals whose red pigment was maximally dispersed, the pigment concentrated to an intermediate condition in contrast to the results obtained with *Orconectes immunis* and *Cambarellus shufeldti* that were described above. When chromatophores were isolated from animals on a white background the degree of red pigment dispersion did not change. The presence of dispersing hormone was first demonstrated by application of extracts to isolated chromatophores. Injection of freshly prepared extracts into intact *Orconectes* always resulted in pigment concentration alone. Fingerman and Aoto (1958b) found that injection of extracts of tissues of *Orco-*

nectes clypeatus that had been kept at room temperatures for 2 hr prior to use caused red pigment dispersion. This result could have been due to activation of an inactive form of the hormone. The effects of long-term adaptation upon the chromatophores were determined with this crayfish also. The results were essentially the same as observed with *Cambarellus*.

Since direct injection of fresh extracts of tissues of *Orconectes* did not produce red pigment dispersion in *Orconectes* it was of interest to determine whether these fresh extracts would disperse red pigment in *Cambarellus*, and vice versa (Fingerman and Lowe 1958). Extracts of central nervous organs as well as eyestalks of *Orconectes* caused concentration and dispersion of red pigment in *Cambarellus*. Extracts of eyestalks of *Cambarellus* caused dispersion as well as concentration of red pigment of *Orconectes*. No dispersion of red pigment in *Orconectes* was caused by central nervous organs of *Cambarellus*. A possible explanation of these data is that *Orconectes* has an excellent feedback mechanism associated with its chromatophore system so that any displacement of red pigment in crayfish on a white background to a more dispersed state is rapidly met with release of red pigment-concentrating hormone which antagonizes the added dispersing hormone. Extracts of the eyestalks of *Cambarellus* alone were able to overcome this mechanism because of the large amount of dispersing hormone in the eyestalk relative to the concentrating substance.

Fingerman and Lowe (1958) also showed that boiling extracts of the supraesophageal ganglia with the circumesophageal connectives attached of *Cambarellus* resulted in potentiation of red pigment-dispersing hormone. Maintenance of extracts at room temperature for 2 hr had the same effect. These results were interpreted on the basis of release of additional hormone from the interior of neurosecretory granules, a mechanism similar to that described by Pérez-González (1957) for the black pigment-dispersing hormone in the sinus gland of *Uca pugilator*.

The connective ganglia of *Cambarellus* and *Orconectes clypeatus* contain more red pigment-concentrating hormone than other portions of the circumesophageal connectives (Fingerman, 1959c). This finding is different from what has been observed with shrimp (Brown, 1946; Knowles, 1953) wherein the tritocerebral commissure contained more chromatophorotropin than did the connective ganglia. Interestingly, of all the portions of the circumesophageal connectives in *Cambarellus* the ganglia alone contain neurosecretory cell bodies (Fingerman and Aoto, 1959). The timecourse of the red pigment-concentrating hormone in the circumesophageal connectives of *Cambarellus* was significantly different from that seen with this hormone in tissues of *Orconectes* when assayed on *Cambarellus*. These hormones may, therefore, be different entities. Electrophoretic analysis at pH 7.2 revealed that the red pigment-dispersing hormone in the eyestalk, supraesophageal ganglia, and circumesophageal connectives of *Orconectes* was electronegative. The substance with the same function in the eyestalks of *Cambarellus* was oppositely charged (Fingerman and Aoto, 1958a). The red pigment-concentrating substance in the supraesophageal ganglia and circumesophageal connectives of both crayfishes was electropositive (Fingerman and Aoto, 1958a; Fingerman, 1959c).

Obviously, the chromatophore systems of *Cambarellus shufeldti* and *Orconectes clypeatus*, the only crayfishes examined in any detail recently, differ from each other considerably. We should, therefore, proceed with caution in transferring information from one species of crayfish to another.

The eyestalks, supraesophageal ganglia, and circumesophageal connectives of the blind cave crayfish *Orconectes pellucidus australis* contain a red pigment-concentrating substance. Assays were performed on *Cambarellus shufeldti* (Fingerman and Mobberly, 1960c). Presumably, this cave dweller that does not possess chromatophores descended from eyed forms that possessed chromatophores as well as mechanisms to regulate migration of their chromatophoral pigments. The persistence of activators of these pigments in blind specimens indicates that either the loss of the controlling mechanism takes longer than the loss of the end organ, or blind forms have given these activators a new function.

CHROMATOPHORES OF STOMATOPODS

The stomatopods are a rather neglected group of crustaceans as far as studies of chromatophores are concerned. Brown (1948b) found that the mantid shrimp, *Chloridella empusa*, shows secondary color changes in response to background. This shrimp is a dark slate color when on a black background and a pale yellow when on white. Specimens from which both eyestalks were removed blanched permanently. However, extracts of these eyestalks had no perceptible effect on eyestalkless *Chloridella* but did darken eyestalkless *Uca* and when injected into eyestalkless *Crangon* strongly darkened the telson and uropods and lightened the remainder of the body.

Knowles (1954) reported that the post-commissure organs of *Squilla mantis* contained large quantities of a substance that dispersed the dark pigment of eyestalkless specimens. In the post-commissure organs of *Squilla*, Knowles (unpublished data, mentioned in Carlisle and Knowles, 1959) has found a substance whose electrophoretic behavior was similar to that of the A-substance. The new material, termed the A'-substance, concentrated the pigment in the white and the large red chromatophores of *Palaemon*, but was without effect on the small red chromatophores. The evidence indicated that the A'-substance may be transformed into the A-substance in extracts kept at room temperature. The A'-substance was presumably the one Knowles (1954) assayed on *Squilla* itself. Obviously much work remains to be done before the chromatophore system of a single species of stomatopod is clearly understood.

CHAPTER 3

CHROMATOPHORES OF INSECTS

FEW species of insects show physiological color changes. Color changes of the dipteran *Chaoborus* (*Corethra*) and the phasmid *Carausius* (*Dixippus*) have been studied in detail. The larvae of *Chaoborus* possess anterior and posterior pairs of tracheal sacs which function as floats to suspend this aquatic insect. Functional melanophores are found on these floats. The melanin disperses when larvae are placed on a black background and concentrates in specimens on a white background. Color changes of other insects were described in the first chapter. Insect chromatophores are not innervated, but rather act under hormonal control or as independent effectors.

In insects, as in crustaceans, chromatophorotropins appear to be neurosecretory products. The brain and the corpora cardiaca form a functionally related group of neuroglandular organs, similar to the medulla terminalis X-organ-sinus gland complex of crustaceans, in which the corpora cardiaca serve as reservoirs for neurosecretory material produced in the brain. Cutting the nerves to the corpora cardiaca results in permanent loss of secretory material from these organs and a considerable increase in neurosecretory material in the brain and in the stubs of the corpora cardiaca nerves attached to the brain. This situation has been observed in the cockroach *Leucophaea maderae* (Scharrer, 1952b), the blowfly *Calliphora erythrocephala* (Thomsen, 1954), and *Carausius morosus* (Dupont-Raabe, 1958).

The similarity of neurosecretory systems among insects, whether or not they possess chromatophores, is intriguing, especially in view of the fact that insects that do not possess chromatophores produce chromatophorotropins (e.g. Brown and Meglitsch, 1940; Gersch, 1956).

Meyer and Pflugfelder (1958) studied neurosecretory cells of *Carausius morosus* by means of an electron microscope. The hormones appear to be produced in the cell bodies and gathered into granules about 100 mμ in diameter.

The mechanism of color change has been investigated in greater detail for *Carausius* (*Dixippus*) *morosus* than for any other insect. However, in 1954 Dupont-Raabe remarked that the mechanisms for control of chromatophores appear to be similar in *Carausius* and *Chaoborus*.

Giersberg (1928) was the first investigator to present a detailed analysis of this process in *Carausius*. He found that high humidity as well as a black background induced darkening. Giersberg also was able to show by an ingenious experiment that this darkening response was mediated by a substance released from an organ in the head. When the abdomen was inserted into a moist chamber, a darkening reaction was initiated in the head and gradually spread toward the posterior end. Darkening was complete in 30–60 min. If a constriction was placed around the thorax, only the portion of the insect anterior to the ligature darkened. Loosening of the ligature soon resulted in darkening of the entire animal. The ligature presumably was not tight enough to prevent passage of nerve impulses. Transection of the nerve cord in the region of the subesophageal ganglion prior to exposure of the posterior end to high humidity prevented the darkening response. However, exposure of the anterior end of these operated animals resulted in darkening of the head and then the rest of the body. Later, Janda (1934) observed that *Carausius* exhibited a 24 hr cycle of color change when maintained in darkness. The insects were dark at night and pale by day. The rhythm could be reversed by illuminating the animals by night and keeping them in darkness by day.

Dupont-Raabe (1951) found that removal of the corpora cardiaca, corpora allata, and frontal ganglion had no effect on color changes in *Carausius*. Removal of the brain, however, always resulted in total cessation of color changes. Brainless animals remained pale. Extracts of brains darkened *Carausius* whereas extracts of corpora allata and frontal ganglion had no effect. Implants of corpora cardiaca, however, produced a slight darkening. Epinephrine and intermedin were ineffective. Three pairs of nerves leave the brain of *Carausius* to innervate the corpora cardiaca. Dupont-Raabe (1956a) feels that chromatophorotropin passes from the brain to the corpora cardiaca through axons of the nervi corporis cardiaca III. This nerve appears to run directly from the tritocerebral neurosecretory cells to the corpora cardiaca. She appears to believe that the tritocerebrum elaborates the darkening principle; the tritocerebrum contains more of this material than any other part of the brain.

Electrophoretic analysis revealed that at pH 7.4–7.8 the chromatophorotropin in the corpora cardiaca of *Carausius* is electropositive (Carlisle, Dupont-Raabe, and Knowles, 1955; Knowles, Carlisle, and Dupont-Raabe, 1955). These investigators felt that this substance may be identical with the A-substance of *Palaemon*. At the same pH range the chromatophorotropin in the brain had an extremely low mobility and was found on both sides of the origin; the authors called this the C-substance. Dupont-Raabe (1956a) postulated that as the C-substance travels to the corpora cardiaca it is transformed into the A-substance.

Dupont-Raabe (1956b) found that the rhythm of color change in *Carausius* depends upon intact connections between the brain and subesophageal ganglion. Perhaps the rhythmical center resides in the subesophageal ganglion as has been postulated by Harker (1960) for the locomotor rhythm of the cockroach *Periplaneta americana*.

Pautsch (1952) showed that head extracts and blood of *Carausius morosus* caused pigment migration in the chromatophores of amphibians (*Rana esculenta, Rana temporaria, Ambystoma mexicanum*), a shrimp (*Crangon crangon*), and an isopod (*Idotea balthica*). His experiments were performed during the day phase of the cycle when *Carausius* would be normally pale. Head extracts and blood always caused the melanin of the amphibians to disperse, but the melanin of *Idotea* assumed an intermediate condition whether it was fully dispersed or fully concentrated at the time of injection. The results obtained with *Crangon* depended on whether blood or head extract was used. Blood induced concentration of pigment in the xanthophores and erythrophores, whereas the pigment in melanophores and leucophores dispersed if concentrated at the start of the experiment but showed a slight concentration if originally dispersed. On the other hand, head extracts always caused pigment concentration in all the chromatophores. The different effects of blood and head extracts on *Crangon* led Pautsch to postulate that at least one chromatophorotropin other than the substances found in the brain and corpora cardiaca normally occurs in the blood of *Carausius*.

Turning to *Chaoborus*, in 1929 Martini and Achundow showed that dispersion of melanin in *C. plumicornis* likewise is controlled by a principle originating in the head. Later, Teissier (1947) noted that the distribution of melanophores in another species of *Chaoborus*, namely *crystallinus*, is the same as in *C. plumicornis*. He noted that the pigment of *C. crystallinus* would disperse in darkness and concentrate in light. This reaction did not appear to be a direct response to illumination but rather seemed to operate through the eyes. If the eyes were covered, the animals behaved as they would in darkness. When a ligature was tied around the middle of the body only the anterior chromatophores continued to show dispersion of melanin. But the posterior end of a ligatured dark animal lightened, presumably because of loss of darkening hormone. The posterior end of a ligatured pale larva remained light. Later Hadorn and Frizzi (1949) found that the brain and subesophageal ganglion of *C. plumicornis* contain a melanin-dispersing substance. In the same year Dupont-Raabe (1949) showed that the brain of *Chaoborus plumicornis* is the major source of darkening hormone. The corpora cardiaca were also very active. Extracts of the corpora allata and frontal ganglion, however, were ineffective. These results were the same as she was to find later with *Carausius* (Dupont-Raabe, 1951). She also noted that brains of phasmids, Saltatoria, culicids, and blattids contained a black pigment-dispersing principle.

Gersch (1956) using a hot needle stimulated various parts of the central nervous systems of *Chaoborus crystallinus* and *C. pallidus* larvae and found that stimulation of the brain caused more darkening than did any other portion of the central nervous system. This finding would be anticipated in view of the results of the other investigators who examined this genus. Gersch also was able to show in assays on *Chaoborus* that the cockroach *Periplaneta americana* produced two chromatophorotropins. He separated these by means of paper chromatography. One, a darkening principle, was found in central nervous organs, the corpora cardiaca, and the corpora allata. He called this material the C-substance, presumably because of its similarity to the C-substance Knowles, Carlisle, and Dupont-Raabe (1955) had found in *Carausius*. He believed the second substance, termed the D-substance, was present in central nervous organs and the corpora cardiaca. The D-substance showed a dilution effect; in dilute solution it caused darkening but in concentrated solutions it caused lightening. Gersch felt that the same two substances were produced by *Chaoborus* but the quantities extracted from this insect were so small that conclusive evidence for them could not be obtained. It should be reiterated, however, that Dupont-Raabe (1949, 1951) had reported that the corpora allata of *Chaoborus* and *Carausius* do not contain a chromatophorotropin, but these glands in *Periplaneta* do and yet cockroaches do not possess chromatophores. Also, the C-substance of *Carausius* occurs in the brain, but not in the corpora cardiaca (Knowles, Carlisle, and Dupont-Raabe, 1955; Carlisle, Knowles, and Dupont-Raabe, 1955), but Gersch found this substance in both the brain and corpora cardiaca of *Periplaneta*.

In 1960, Gersch, Fischer, Unger, and Koch reported the results of further efforts to purify the C and D factors. In assays on *Carausius* they found that the C and D factors of Gersch (1956) were actually four substances which they called C_1, C_2, D_1, and D_2. They found that C_1 and D_1 were the only fractions that were chromatophorotropic. The effects of the chromatophorotropins on *Carausius* were the same as Gersch (1956) had observed with

Chaoborus. The infrared spectrum of the D_1 substance indicated it was a peptide. These investigators were able to crystallize both C_1 and D_1. A concentration of 10^{-10} g/ml or greater of D_1 caused significant lightening; darkening occurred with a concentration of 10^{-11} g/ml. They had started with 3200 *Periplaneta* and obtained about 50 gamma of crystalline D_1-substance.

Through use of alcohol-extraction techniques, Rounds and McClain (1961) were able to show that central nervous organs of the cockroach *Blaberus giganteus* contain pigment-concentrating and pigment-dispersing principles. Their assays were performed on a "crustacean". It would be interesting to learn whether these chromatophorotropins are the C_1- and D_1-substances of Gersch, Fischer, Unger, and Koch (1960).

Mothes (1960) described the results of some very interesting experiments dealing with color changes in *Carausius morosus.* Assays of portions of the brain revealed that the D_1-substance is formed in the pars intercerebralis. The C_1-substance was restricted to a portion containing both the deutocerebrum and tritocerebrum. It will be recalled that Dupont-Raabe (1956a) had stated that she believes the darkening substance of *Carausius,* her C-substance, is formed in the tritocerebrum. Mothes also found the C_1-substance in the corpora cardiaca and corpora allata; D_1-substance in the corpora cardiaca but not in the corpora allata. He detected a daily cycle of hormone titer in the blood and tissues. During the night, when the animals were normally darker because of their rhythm of color change, the amounts of both C_1- and D_1-substances were greater in the blood than during the day. However, during each 24 hr, the relative amounts of the two substances changed, so that the color of the animal would appear to depend upon the relative amounts of these substances as well as their absolute quantities, just as had been postulated for *Cambarellus* by Fingerman and Aoto (1958a). During the night the amounts of both substances present in the subesophageal ganglion and abdominal nerve cord of *Carausius* were greater than during the day-time. The quantities in the brain, however, were greater by day than at night. The quantities in the corpora allata and corpora cardiaca remained constant at all times of day and night. These cycles of hormone titer could be shifted 12 hr out of phase by reversed illumination. The brain rhythm was the most labile. An exciting observation of his was that the chromatophores of intact animals would respond to darkening substance at night only when this animal because of its color change rhythm would be darkening. He believes the lack of response by day was due to a cycle in the hypodermal cells themselves. However, isolated chromatophores responded during the day. Mothes postulated that this response was due to a heightened sensitivity of the chromatophores *in vitro*. Further investigation of this phenomenon certainly should be carried out. It is surprising that this cycle of responsiveness was not noticed by the earlier investigators.

Key and Day (1954b) reported that the chromatophores of the grasshopper *Kosciuscola tristis* are independent effectors. No nervous or hormonal control of the chromatophores was apparent. Chromatophores on pieces of isolated integument functioned perfectly well. The chromatophores appeared to respond only to temperature changes (see Chapter 1). No response to black or white backgrounds, relative humidity, or crowding was detected.

Within the past decade some very interesting experiments on morphological color change in insects have been performed. These will be mentioned briefly since the emphasis herein is on physiological color changes. The orthopteran *Acrida turrita* shows a morphological color change at the time of molting (Ergene, 1950). The larvae have two color phases, green and yellow, and form the appropriate pigment to match their particular background. Joly (1951, 1952) showed an endocrine basis for morphological color changes. He found that the corpora allata are necessary for the formation of green pigment in *Locusta migratoria* and *Acrida turrita*. Otherwise these orthopterans remain pale yellow. In 1952 Ergene reported the eyes are needed for the yellow form to became green. However, blinded green larvae became yellow even when maintained on a green background. He (Ergene, 1953, 1954, 1955a, b, c, d, 1956a, b, c) reported similar results with *Mantis religiosa*, *Oedaleus decorus*, and *Tylopsis liliifolia*.

When early nymphs of *Mantis religiosa*, having green hemolymph, were reared in yellow cages, their hemolymph color changed to yellow before the integument color became established. Similarly, when nymphs with yellow hemolymph were reared in green cages, their hemolymph turned green. When given an experimental "choice" of background 76% of the yellow long-horned locusts *Tylopsis liliifolia* were found on a yellow background after 18 hr, the same percentage of green specimens on a green background. Morphological color changes in insects appear to depend upon background alone and not on temperature, light intensity, or humidity. Girardie (1962) found that the pars intercerebralis of *Locusta migratoria* inhibits the principle from the corpora allata that causes the formation of green pigment.

CHAPTER 4

CHROMATOPHORES OF CEPHALOPODS

CHROMATOPHORES of cephalopods are controlled primarily by nerves. Blood-borne substances have a secondary role. The structure and colors of these chromatophores were described earlier. Color changes in cephalopods are rapid. Quite commonly color changes pass over the body from one end to the other, so that one end may be dark while the opposite end is still pale. However, when animals are placed on a particular background, they tend to maintain their adaptation to that shade (Kühn, 1950). The eyes are the main sense organs sending stimuli to the color change centers. When one eye is removed, the entire animal continues to change color, but on the blinded side the changes are not as striking (Sereni, 1930a). Complete blinding further reduces but does not abolish color changes. Tactile stimulation, such as stimulation of the suckers on the tentacles, may result in a color display. Postural changes, probably involving tactile and equilibrium centers, also contribute significantly to color changes. For example, the chromatophores on the functional ventral surface are normally more concentrated than on the dorsal surface. Ventral illumination does not reverse this condition but turning an animal over will.

Sereni (1930a) obtained evidence for the existence of three chromatic centers in the nervous system. One, situated in the supraesophageal ganglia, is a motor center and consists of two portions, one for each side of the body. These motor centers can not substitute for each other. The second, a general color center in the central ganglia, controls the first one. The third is an inhibitory center located in the cerebral ganglia. The general color center and inhibitory center are also symmetrically placed, but each portion may act for the opposite side of the body as well

93

as for its own. Evidence for excitatory and inhibitory fibers innervating each chromatophore has been presented by Sereni (1928,
1930a). Bozler (1928) has also presented evidence for dual innervation of cephalopod chromatophores. However, he believed the
nerve fibers were tonic and tetanic. The muscle fibers of the chromatophores appeared to contain two types of myofibrils, peripheral coarse ones and central fine ones. In 1931 Bozler reported
these muscle cells can exhibit both tetanic and tonic contractions.
Furthermore, the tetanic contractions appeared to be carried out
by the peripheral coarse myofibrils and the tonic contractions by
the thin central fibrils.

Substances that appear to be hormones operate primarily by
influencing the nervous centers and secondarily by acting directly
on the chromatophores. Tyramine, for one, has been found in
the blood. This material, produced by the posterior salivary
glands, appears to cause darkening primarily by increasing the
tonus of the motor centers (Sereni, 1929a, b, c, 1930b, c). Extirpation of the posterior salivary glands results in paler specimens
and loss of chromatophore tone. Tyramine is more concentrated
in the blood of dark species of *Octopus* (e. g. *macropus*) than of
light species (*O. vulgaris*). Furthermore, blood from *O. macropus*
will darken *O. vulgaris*. However, blood from *O. vulgaris* will
blanch *O. macropus*, suggesting the presence of a lightening substance. Betaine is the lightening-substance found in cephalopods
that appears to be involved in endocrine control of the chromatophores. This principle causes paling (decreases the chromatophore tone), presumably by stimulating the inhibitory center.

Evidence that tyramine and betaine may operate secondarily
directly on the chromatophores comes from studies on portions
of denervated skin. Both substances under such conditions cause
chromatophore dispersion. The opposite actions of betaine, i. e.
blanching as a result of stimulation of the inhibitory center in
intact specimens and darkening by direct stimulation of isolated
chromatophores, certainly deserves further consideration.

More recently, Erspamer and Asero (1953) found large quantities of serotonin in the posterior salivary glands of *O. vulgaris*.

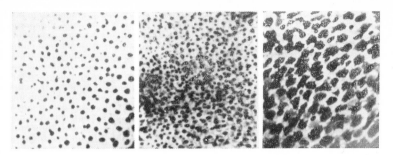

FIG. 26. Chromatophores in the skin of *Octopus vulgaris*. Left, control, pigment in an intermediate state under cold narcosis. Center, pigment concentration after exposure to serotonin. Right, pigment dispersion after exposure to acetyl choline. (From Kahr, 1959.)

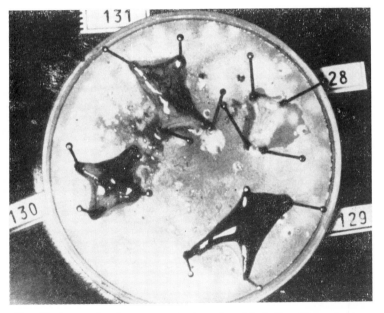

FIG. 27. Antagonism between serotonin and acetyl choline. Piece of skin 128 from *Octopus* was exposed to 10^{-5} mg. serotonin alone. The other pieces were exposed to the same concentration of serotonin and in addition 131 had 10^{-8} mg. acetyl choline, 130 had 10^{-7} mg., and 129 had 10^{-6} mg. (From Kahr, 1959.)

Smaller quantities were noted in the blood. Kahr (1959) attempted to determine the role of serotonin in the chromatics of the same cephalopod. He found that serotonin concentrated all of the chromatophoral pigments. He also reported that histamine had no effect on the chromatophores but acetyl choline was a potent chromatophore-expanding agent whose action was antagonized by serotonin (Figs. 26, 27). The actions of these drugs were the same whether intact specimens or pieces of skin were used for the assays. The roles of histamine and acetyl choline described by Kahr were different from those reported by Sereni (1930b, c) for the same animal. Sereni had stated that histamine caused darkening of intact specimens whereas acetyl choline caused them to blanch. Obviously much remains to be learned about the control of color changes in this groups of organisms, especially the role of blood-borne substances. Also, modern electrophysiological techniques should yield very interesting information concerning the dual innervation of these chromatophores.

CHAPTER 5

CHROMATOPHORES OF ECHINODERMS

LONG ago Von Uexküll (1896) described the color changes of the Mediterranean sea urchins *Arbacia pustulosa* and *Centrostephanus longispinus*. Both organisms darkened when illuminated. This observation has been confirmed by Kleinholz (1938b) *Arbacia punctulata*, an American species, shows no color change (Parker, 1931).

More recently, Millott (1950, 1952) described the color changes of the sea urchin *Diadema antillarum*. The urchins would become lighter at night, certain parts of the test becoming completely white as the melanin concentrated. Illumination would again cause darkening. During the intermediate phases of the color change a pattern appeared which was a brilliant blue due to dispersion effects caused by the receding black pigment. About 90 min were required for the color change, which was more marked in young forms than in adults. The chromatophores appeared to operate independently of the central nervous system. Millott also noted a daily rhythm of color change; animals kept in darkness were slightly darker by day than at night. This observation is interesting in view of the facts that (1), the chromatophores do not appear to be under central nervous control and (2), a direct relationship exists between depth of color and light intensity.

Yoshida (1956) felt that since Millott (1950, 1952) had illuminated large areas of *Diadema antillarum*, the possibility existed that peripheral nerves were stimulated by the illumination. He, therefore, performed some experiments using a minute light spot on the test of *Diadema setosum*. When a pigmented part of a chromatophore was illuminated, the pigment in that part could not be withdrawn from the illuminated area. When a pigment-free area of a chromatophore was illuminated, the pigment dispersed to cover the illuminated area. He concluded that photosensitivity resided in the chromatophore itself, probably in the hyaloplasm.

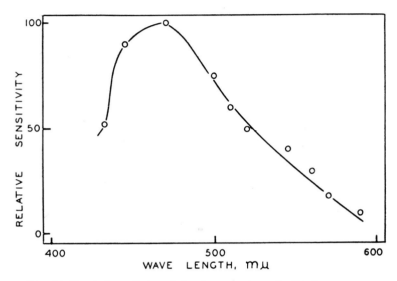

FIG. 28. Spectral sensitivity of the chromatophores in *Diadema setosum*. (Redrawn from Yoshida, 1957.)

In view of the direct response to light of the chromatophores on *Diadema setosum*, Yoshida (1957) thought it would be interesting to determine the spectral sensitivity of these cells. Chromatophores on the aboral region of the interambulacra were used, since these cells are more responsive than those in other regions of the body. The relative threshold energy was determined for each of a series of wave lengths of monochromatic lights. Reciprocals of the relative energies were considered an expression of the relative sensitivity of the chromatophores. The wave length of 468 mμ was most effective (Fig. 28). The sea urchins *Psammechinus miliaris* and *Diadema antillarum* showed spine movements that were initiated by changes in light intensity (Millott and Yoshida, 1956, 1957). The spectral response for the shadow reflex in both species was similar to that of the chromatophores in *Diadema setosum*. Perhaps a common pigment is involved in both responses. Obviously, much research must be done before we understand fully the mechanism of color change in echinoderms. For example, it would be interesting to learn what light does to the chromatophores to cause pigment migration.

Unger (1960) reported the presence of neurosecretory cells in the nerve ring and radial nerves of the starfish, *Marthasterias glacialis*. Through use of filter paper chromatography he was able to separate two bluish fluorescent substances (R_F 0·19 and R_F 0·33) from extracts of the nervous tissue. He furthermore found that when isolated pieces of skin from the aboral surface were exposed to the faster moving fraction for 10 hr a darkening reaction occurred whereas the other component caused blanching. These observations will certainly stimulate other investigators of echinoderm physiology. Unger's paper appears to constitute the first report of color changes in starfishes as well as of neurosecretory cells in echinoderms.

CHAPTER 6

CHROMATOPHORES OF VERTEBRATES

TRUE chromatophores occur among the cold-blooded vertebrates, namely, the cyclostomes, the elasmobranchs, the teleosts, the amphibians, and the reptiles. The mechanisms of control are varied, including instances of hormones alone (aneuronic chromatophores), nerves alone, and both hormones and nerves. Furthermore, chromatophores may be singly (mononeuronic chromatophores) or doubly (dineuronic chromatophores) innervated. With regard to endocrines, in some cases where hormones have been implicated excellent evidence exists for pigment-concentrating and pigment-dispersing substances, whereas in other instances the evidence is good for a darkening material alone.

SOURCES OF CHROMATOPHOROTROPINS

The secretory products of several organs have been shown to possess chromatophorotropic activity. These organs comprise the pituitary gland, pineal gland, adrenal gland, thyroid gland, and even the skin itself. Pickford and Atz (1957) discussed the morphology and physiology of the pituitary gland in vertebrates. Their emphasis, however, was on fishes. A more recent discussion of this subject was presented by Green and Maxwell (1959). The pituitary gland of all vertebrates is basically similar. Embryologically, this gland is derived from two components, the adenohypophysis and the neurohypophysis. The former develops from an ectodermal pouch in the roof of the mouth; the latter forms from an outpouching of the floor of the diencephalon, the infundibulum. The hypophyseal pouch from which the adenohypophysis develops usually closes over in adult life, but in some fishes, most commonly the cyclostomes, it remains open. The terminology of the portions of the adenohypophysis in fishes has become very confused because many authors gave different names to the same part of the organ as a result of attempts, based on insufficient information and misconceptions, to homologize the parts of the adenohypophysis in fishes with the rather clear-cut portions of the pituitary gland in higher classes of vertebrates. Pickford and Atz (1957) attempted to eliminate this confusion to some degree with a new system of describing the parts of the pituitary gland of fishes. This system has merit. It is unfortunate that such a scheme was not adopted before the confusion became so great. Pro-adenohypophysis was suggested for pars anterior, pars follicularis, or anterior glandular region; meso-adenohypophysis for transitional lobe, Übergangsteil, or middle glandular region; and meta-adenohypophysis for pars intermedia or posterior glandular region. The portions of the adenohypophysis in higher vertebrates are the pars distalis which consists of the main mass of secretory cells, the pars tuberalis which is an upgrowth around the stalk of the infundibulum, and the pars intermedia which tends to fuse with the neurohypophysis.

The morphology of the hypophysis and hypothalamic neurosecretory system has been described for a few fishes since the appearance of the monograph by Pickford and Atz (1957). Among such is the description of these structures in the goby *Lepidogobius lepidus* by Kobayashi, Ishii, and Gorbman (1959). The pituitary gland of this fish is interesting in that it is located in a hollowed out space within the brain instead of being at the end of an infundibular stalk. The infundibular region consequently is concave with corresponding alterations in the pattern of neurosecretory fibers. The largest cells of the adenohypophysis are in the meta-adenohypophysis, this region is also more vascular than any other part of the adenohypophysis. Penetration of the meta-adenohypophysis of this fish by stainable neurosecretory fibers was observed in a few glands. Oztan and Gorbman (1960) described these structures in two larval lampreys, *Petromyzon marinus* and *Entosphenus lamottei*. Many neurosecretory cells were observed in the hypothalamus; their axons led to the neurohypophysis, as has been reported previously for several other vertebrates (Green, 1951; Scharrer and Scharrer, 1954). With respect to amphibian neurosecretory systems, Jørgensen, Larsen, Rosenkilde, and Wingstrand (1960) determined the effects of extirpation of the median eminence on the functioning of the pars distalis in the toad *Bufo bufo*. They found that only when the cut ends of the hypothalamic tracts normally terminating in the median eminence made contact with the pars distalis did it resume functioning. The suggestion was made that the pars distalis activating substances that are normally liberated in the median eminence are produced in hypothalamic centers as neurosecretory products.

Recent studies on vertebrate neurosecretory systems in which the electron microscope has been employed have emphasized the neurosecretory granules in the pituitary stalk and the neurohypophysis (Fujita, 1957; Green and Maxwell, 1959; Bargmann, 1958; Duncan, 1956; Palay, 1955). In some species the neurosecretory granules appear to be bounded by a membrane and in some species no membranes have been observed. In still other instances some axons appear to contain granules without membranes while neighboring axons contain granules surrounded by a distinct membrane. The significance of these membranes is not clear. Dawson (1953) presented some inconclusive evidence for the termination in the pars intermedia of neurosecretory fibers that originate in the pars nervosa of the frog *Rana pipiens*.

Hormone-containing granules have been isolated from the adrenal medulla of cattle. These granules account for 80 to 90% of the epinephrine and norepinephrine in the gland (Hillarp, Lagersted, and Nilson, 1953; Blaschko and Welch, 1953). Hillarp and Nilson (1954) and Blaschko, Hagen, and Welch (1955) studied these catechol amine-containing granules and obtained strong evidence that the granules are bounded by a semipermeable membrane of lipid or lipoprotein nature. Furthermore, these granules, which are stable in isotonic solutions of saline or sucrose, release their amines when exposed to agents known to disrupt biological membranes. These granules, however, do not appear to be true secretory granules, i.e. they do not appear to leave the cells during secretion, but rather the stored catechol amines seem to be released from the granules before leaving the cell (Carlsson, Hillarp, and Hökfelt, 1957). One would certainly like to know if this situation is true of granules in neurohaemal organs.

CHEMISTRY OF CHROMATOPHOROTROPINS

The comparative biochemistry of protein and polypeptide hormones has been discussed recently by Geschwind (1959) and Lerner (1961). Two general types of intermedin have been found and named α-MSH (melanophore-stimulating hormone) and β-MSH. Alpha-MSH is a very basic peptide with an isoelectric point between 10.5 and 11.0 (Lee and Lerner, 1956). Its amino acid sequence was determined by Harris and Lerner (1957) from porcine pituitaries. The structure is Acetyl-ser-tyr-ser-met-glu-his-phe-arg-try-gly-lys-pro-val-NH_2. Alpha-MSH from beef, horse, and monkey pituitaries has the same structure as porcine α-MSH.

Beta-MSH exists as an acidic or neutral peptide. The structures of porcine, bovine, and sheep β-MSH were originally worked out by two groups of investigators working independently (Harris and Roos, 1956; Geschwind, Li, and Barnafi, 1956, 1957a, b; and Geschwind and Li, 1957). The molecule is a chain of 18 amino acids: H-asp-X-gly-pro-tyr-lys-met-glu-his-phe-arg-try-gly-ser-pro-pro-lys-asp-OH. The X may be glutamic acid or serine. The substitution of an acidic amino acid (glutamic acid) in β-MSH for neutral serine explains the difference between the respective isoelectric points of 5.8 and 7.0. It had originally been thought that β-glutamyl-MSH occurred in hog pituitaries, β-seryl-MSH in beef pituitaries, and that sheep pituitaries possessed both types. However, Burgers (1961) devised electrophoresis and assay techniques that enabled him to assay MSH from single pituitaries. His assays were done on the lizard *Anolis carolinensis*. He found that beef, pig, and sheep intermediate lobes contain three MSH components that behave electrophoretically like α-MSH, β-glutamyl-MSH, and β-seryl-MSH. The three components were present in each single pituitary gland in a ratio that appeared to be more or less constant within a species and which was characteristic for that species. He feels that the reason others did not find these in the same gland was due to differences in extraction methods and bioassay techniques.

Harris (1959) determined the structure of human β-MSH. It is H-ala-glu-lys-lys-asp-glu-gly-pro-tyr-arg-met-glu-his-phe-arg-try-gly-ser-pro-pro-lys-asp-OH, a molecule basically similar to ox and pig β-MSH, but containing four extra amino acids.

Corticotropin from the anterior lobe of the pituitary also causes darkening by dispersing the pigment in melanophores. Johnsson and Högberg (1952) reported that MSH and adreno-corticotropic hormone darkened *Rana temporaria*. They concluded these substances were closely related or identical. Geschwind, Reinhardt, and Li (1952) showed that these substances were not the same. Morris (1952), Karkum, Kar, and Mukerji (1953), Ketterer and Remilton (1954a), and Forgács (1956) came to the same conclusion as Geschwind, Reinhardt, and Li (1952). As an example of the type of evidence obtained to support the idea that they are different substances is the fact that heating in alkali potentiates MSH but destroys ACTH. The chemical structures of corticotropins from sheep, ox, and pig pituitaries have been determined (Howard, Shepherd, Eigner, Davies, and Bell, 1955; White and Landmann, 1955; Li, Geschwind, Cole, Raacki, Harris, and Dixon, 1955; Shepherd, Willson, Howard, Bell, Davies, Eigner, and Shakespeare, 1956; and Li, Dixon, and Chung, 1958). Each molecule consists of a chain of 39 amino acids. However, only 24 are necessary for the molecule to be active. The sequence of these 24 is the same in sheep, ox, and hog pituitaries. Species differences among the molecules show up in the remaining 15 amino acids. In sheep and ox corticotropins the 15 amino acid fragments have the same amino acid composition but differ in the sequence. The sequence of amino acids in the active 24 amino acid fragment is: ser-tyr-ser-met-glu-his-phe-arg-try-gly-lys-pro-val-gly-lys-lys-arg-arg-pro-val-lys-val-tyr-pro. It is interesting to note that in α-MSH, β-MSH, and corticotropin there is a tyrosine followed by either a serine (α-MSH, corticotropin), lysine (pig, beef, sheep, horse β-MSH), or arginine (human, monkey β-MSH) and then a heptapeptide (met-glu-his-phe-arg-try-gly) followed by a serine (all β-MSH) or lysine (α-MSH, corticotropins), and then by a proline. However, there is no evidence that any intermedin has corticotropic activity. Perhaps this similarity of amino acid sequences is the basis for the fact that all of these molecules have a melanin-dispersing action.

Schwyzer and Li (1958) were able to synthesize the pentapeptide l-histidyl-l-phenylalanyl-l-arginyl-l-tryptophyl-glycine. This fragment, important because it occurs in ACTH and the intermedins, will cause melanin dispersion. Boiling in 0.1N NaOH for 15 min will potentiate the action of MSH and of this pentapeptide. Potentiation appears to be due to partial conversion of the arginine residue to ornithine.

Burgers (1956) found that epinephrine darkened the African clawed toad, *Xenopus laevis*. He studied the relationship between the chemical structure and melanophore activity of several analogs of epinephrine and found that a hydroxyl group must be in the three and four position of the phenyl nucleus for a positive effect. The presence of a hydroxyl group at the 1′C atom on the side chain is also important. Substitution of a hydrogen atom for this hydroxyl group decreased the activity; a methoxy group eliminated it.

Lerner, Case, Takahashi, Lee, and Mori (1958) reported the isolation of a material, which they called melatonin, from beef pineal glands. This substance is a very potent melanin-concentrating agent for *Rana pipiens* and is able to prevent darkening by MSH. It was not found in the beef pituitary or hypothalamus. Melatonin has no epinephrine or norepinephrine-like activity on the rat uterus or serotonin-like activity on clam heart. Its structure is N-acetyl-5-methoxytryptamine (Lerner, Case, and Heinzelman, 1959; Lerner, Case, and Takahashi, 1960). Melatonin is 100,000 times more potent than norepinephrine in causing melanin concentration. In a concentration as low as 10^{-12} g/ml, melatonin can prevent and partially reverse the darkening actions of ACTH and MSH.

Imai (1958) determined some of the chemical properties of the melanin-concentrating hormone occurring in the pituitaries of the carp, *Cyprinus carpio*, and the catfish, *Parasilurus asotus*. Enami (1955) had originally discovered the existence of this principle. The substance appeared to be digested by trypsin but not by pepsin and was soluble in acetone but not in methanol, ethanol, ethyl ether, or chloroform.

CHROMATOPHORES OF CYCLOSTOMES

Control of chromatophores in cyclostomes has not received much attention. Young (1935) analyzed in detail the color changes of larval and adult specimens of *Lampetra planeri*. His publication represents the most detailed report of research that has been done on chromatophores of cyclostomes. He observed a rhythm of color change whereby the fish were dark by day and light at night. This rhythm appeared to be determined primarily by changes in the environmental light intensity. Specimens would remain dark as long as they were illuminated. However, some of the specimens continued to exhibit a 24 hr cycle of color change in constant darkness. The rhythm may depend in part upon an endogenous component. No background response was noted. A slight temperature effect was detected, however. Low temperatures favored paling. The melanophores were found to be aneuronic. Removal of the pituitary from the ammocoetes larva resulted in concentration of the pigment in the dermal melanophores. Mammalian pituitary extracts caused darkening. Removal of the pineal and parapineal organs from ammocoetes larvae resulted in permanent melanin dispersion and loss of the rhythm of color changes. Young concluded that the pineal eye is a light receptor which sends impulses to the pituitary to inhibit release of darkening hormone. The same results were obtained when the pineal organ and lateral eyes were removed from adults. It would be interesting to learn if a melatonin-like material is actually involved. In the same year, Young and Bellerby (1935) reported that repeated injections of extracts of anterior lobes from ox pituitaries caused a slight concentration of melanin in larval *Lampetra planeri*. This reaction may have been a non-specific response or due to a melanin-concentrating material. Coonfield (1940) found that *Myxine glutinosa* would respond to black and to white backgrounds requiring 2 or 3 days for the process. Epinephrine gradually blanched this cyclostome. More recently Lanzing (1954) found that extracts of the pituitary and the adjacent region of the brain of *Lampetra fluviatilis* darkened the frog *Rana esculenta*.

CHROMATOPHORES OF ELASMOBRANCHS

The ability to change color is not shared by all elasmobranchs. These fishes can be separated into two groups on the basis of their ability to exhibit a background response. One group requires hours or even a few days to exhibit a background response, quite slow compared with some teleosts. In this group belong *Mustelus canis* (Parker and Porter, 1934), *Raja brachiura*, *Raja maculata*, *Rhina squatina*, *Scyllium canicula*, and *Scyllium catulus* (Hogben, 1936), *Trigon pastinaca* and *Raja undulata* (Vilter, 1937), and *Squalus acanthias* (Parker, 1936a, 1937; Waring, 1938). The second group shows no background adaptation or, at best, limited responses that became apparent only after days or weeks of exposure to a particular shade of background. Examples of this group are *Raja clavata*, *Raja batis* (Schaefer, 1921), *Urolophus testaceus* and *Trygonorrhina fasciata* (Griffiths, 1936) and *Torpedo marmorata* (Veil and May, 1937). Removal of the entire pituitary gland or just the neurohypophysis plus the meta-adenohypophysis results in concentration of pigment in elasmobranch melanophores as shown by Lundstrom and Bard (1932) with *Mustelus canis*; Lewis and Butcher (1936) with *Squalus acanthias*; Hogben (1936) with *Scyllium*, *Raja brachiura*, *Raja maculata*, *Raja squatina*, *Scyllium catulus*, *Raja clavata*, and *Raja microcelata*; Veil and May (1937) with *Torpedo marmorata*; Vilter (1937) with *Trigon pastinaca* and *Raja undulata*, and Abramowitz (1939) with *Raja erinacea*.

Although there is universal agreement that darkening in elas-mobranchs is caused by a secretion from the pituitary gland, no such agreement exists concerning the mode of lightening in this group of fishes. No less than four theories have been proposed to explain this phenomenon. The first and simplest theory is the "Unihormonal Theory" (Lundstrom and Bard, 1932). This theory states that melanin concentration is due merely to removal of intermedin from the blood.

Another theory, one of the "Nervous Theories", was proposed by Parker and Porter (1934) and Parker (1935a, b) on the basis of experiments with the smooth dogfish *Mustelus canis*. According-ing to this theory paling is due to innervation of the melanopho-res by pigment-concentrating fibers. Evidence in favor of this concept was obtained by making incisions in the fins or body proper which gave rise to light bands or splotches which persisted in some cases for several days (Fig. 29). Melanin concentration is supposed to be due to release of a mediator substance from the endings of the nerve fibers which had been stimulated by cutting. Parker (1935a) had found that electrical stimulation of fins had the same blanching effect as did cutting. In the same year Parker (1935b) reported that he obtained a substance by extracting fins of *Mustelus* which when injected into dark dogfish caused melanin concentration in the vicinity of the injection. This substance is soluble in oil but not in water. He believed it was the mediator substance produced by the pigment-concentra-ting nerves. This substance is now called selachine (Parker, 1942). Parker (1936b) found that the melanophore system of new-born pups of *Mustelus canis* responds in the same way as the system of the adults. In support of the "Nervous Theory" Parker (1936c) showed that if a second cut in a fin distal to the first was made after the effects of the first cut had disappeared, a second light band would appear, probably due to restimulation of the pig-ment-concentrating nerve. Parker (1936a) reported that not all elasmobranchs respond to cuts as does *Mustelus*. In *Squalus acanthias* light bands either failed to appear or were extremely faint. No trace of them was observed in *Raja erinacea*. Young

(1933), Waring (1936), and Wykes (1936) could find no trace of bands in several other species of elasmobranchs.

Still another theory, proposed by Hogben (1936) and Waring (1938), states that blanching is due to release of a substance from the pro-adenohypophysis of the elasmobranch pituitary (Dualistic Hormone Theory). This principle they termed W-substance or whitening substance. After removal of the pro-adenohypophysis the fishes remain permanently dark and cannot respond to a white background. This was shown by Hogben (1936) for *Scyllium canicula*, *Scyllium catulus*, and *Raja brachiura*; by Waring (1938) for *Squalus acanthias*; and by Abramowitz (1939) for *Mustelus canis* and *Raja erinacea*. Hogben (1936) and Waring (1938) interpreted these findings on the basis of loss of pituitary lightening hormone whereas Abramowitz (1939) felt these results were due to operative brain damage. However, implants of the pro-adenohypophysis had no significant effect on the animal's color (Waring, 1938) nor did blood transfused from a light animal into a dark animal have an effect (Parker and Porter, 1934; Parker, 1937). But, when intermedin was injected into hypophysectomized dogfishes (Waring, 1938) the response was greater than when injected into fishes from which the neurohypophysis and meta-adenohypophysis but not the pro-adenohypophysis had been removed.

The remaining theory, the second "Nervous Theory", was proposed by Vilter (1937) who felt that blanching is due to tonic action of the sympathetic nervous system allowed expression when the influence of the pituitary is removed or diminished. This theory was based on his studies with *Trigon pastinacea*, *Raja undulata*, and *Torpedo marmorata*.

Abramowitz (1939) considered the several theories of the mechanism of paling in elasmobranchs and concluded that the "Unihormonal Theory" was the only one that could explain all the results obtained by the several investigators of this problem. He considered the evidence for the W-substance at best circumstantial. The "Nervous Theories" of Parker and Vilter were criticized mainly on the basis of the fact that there is no anatomical evidence that chromatophores of elasmobranchs are innervated. This question of innervation could easily be clarified with the excellent nerve stains that have been developed recently. The theory of Parker was also criticized on the basis that the cuts he made in the skin could have interfered with the circulation of the fin and that the paling actually resulted from lack of intermedin. However, Parker (1938a) had already shown that cutting off the blood supply to a region of the body of *Mustelus canis* did not interfere with the lightening response when cuts in the skin were made. Another project well worth undertaking would be a survey of elasmobranchs to determine whether a single mechanism for paling evolved in the elasmobranchs or whether the group is as diverse as the four theories would indicate.

CHROMATOPHORES OF TELEOSTS

The control mechanisms for teleost chromatophores present an even broader spectrum of types than has been postulated in the elasmobranchs. Teleost fishes have been separated into two groups on the basis of their response to pituitary extracts (Pickford and Atz, 1957). In the first group the response is always melanin dispersion. Teleosts of the second group are those in which implants and crude extracts of the pituitary induce melanin concentration but in which highly purified intermedin has little or no dispersing effect. Presumably, in such teleosts color changes are primarily under nervous control.

There is general agreement that teleost melanophores are innervated by at least a pigment-concentrating axon. Evidence for this belief has been obtained by many investigators from experiments involving electric stimulation of nerves, nerve tracts, and nerve centers. Stimulation of these structures results in melanin concentration in part or all of the animal, depending upon the experimental conditions. However, if prior to stimulation, a part of the skin is denervated, blanching will not occur in that region. The evidence that has been obtained for pigment-dispersing fibers has usually been indirect and does not seem as convincing to some investigators as the evidence for pigment-concentrating fibers. Parker (1943, 1948) believes that the concentrating fibers belong to the sympathetic division of the autonomic nervous system (adrenergic), the dispersing fibers to the parasympathetic division (cholinergic).

The chromatophore systems of the killifish *Fundulus*, the eel *Anguilla*, the catfish *Ameiurus*, and the minnow *Phoxinus* have been worked out in greater detail than for other teleosts. The color change systems of these fishes will be discussed in turn, followed by a consideration of recent research with other fishes.

That the pituitary gland of some fishes is not necessary for color changes is obvious from the experiments of Matthews (1933) and Abramowitz (1937b) who noted that hypophysectomized *Fundulus heroclitus* were still able to respond to changes in background about as well as intact specimens. Presumably, the responses in hypophysectomized specimens depend primarily upon an intact nerve supply to the chromatophores. Abramowitz (1937b), however, found enough MSH in the blood of *Fundulus* to suggest to him that this hormone plays a significant, even if slight, role in the chromatic physiology of *Fundulus*. This point needs further study in view of the fact that extracts of *Fundulus* pituitaries are ineffective in intact *Fundulus* (Kleinholz, 1935; Abramowitz, 1937b) but will darken a blanched, denervated portion of the caudal fin, showing that when the melanophores of *Fundulus* are released from nervous control, they will respond to MSH (Kleinholz, 1935).

Mills (1932a, b, c, d) had previously presented evidence for dual innervation of the melanophores in *Fundulus*. Denervated melanophores did not respond to a change in shade of background as rapidly as those with an intact nerve supply. This observation indicated to her that a mediator substance was released from the intact terminals, acted first on the melanophores these fibers innervated, and then diffused into the denervated chromatophores. When melanophore nerves in the tail fin were cut, the resulting areas that remained light when the fish was on a black background, or dark when the fish was on a white one, did not always coincide. Mills felt this experiment showed neigh-

boring dispersing and concentrating axons do not always innervate the same melanophores. She was also able to stimulate the
concentrating axons electrically.

Study of caudal bands, produced by making short transverse
cuts near the root of the caudal fin, enabled Parker to analyze
the chromatophore system of *Fundulus* in even greater detail than
did Mills (Parker, 1933, 1934a). Such an incision would denervate
an area that extended from the cut to the posterior margin of the
tail. After the cut was made in a fish on a white background, the
pigment in melanophores of the denervated area dispersed and

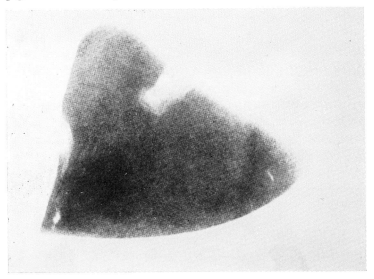

FIG. 29. Dorsal view of the left pectoral fin from a darkish *Mustelus* showing a pale band caused by a transverse cut. The band is believed to have
resulted from the action of a concentrating material secreted from nerves
stimulated by the cutting. (From Parker, 1937.)

was apparent as a dark stripe running the length of the tail (Fig. 30).
This dark band gradually faded in a fish on a white background.
If this fish was transferred to a black container, the body darkened. The stripe, however, would stand out as a light one but
would gradually darken to match the rest of the fish. If this dark
fish were returned to a light container, the body lightened but

the band would stand out as dark and once again gradually fade. The formation of the dark caudal band was interpreted as evidence for a melanin-dispersing nerve fiber. Cutting this fiber stimulated it into activity.

He explained the slow darkening and lightening of denervated caudal bands as due to diffusion of mediator substances into the caudal band from adjoining regions. If after a caudal band had blanched a new transverse cut was made within the area of the old band but slightly distal to the old cut, a second dark band would appear reaching from the new cut over a part of the old band to the edge of the tail. Such a revival of an old band showed that the melanin-dispersing nerve fibers severed originally were still capable of activity.

Parker and Porter (1937) found that maintenance of *Fundulus* on a white background for periods up to 32 hr caused a progressive hastening of the dark-to-pale response and slowing of the pale-to-dark one. Residence in a black container hastened the pale-to-dark response and slowed the dark-to-pale one. These results were interpreted in terms of quantities of mediator substances secreted by the nerve terminals. In the case of a *Fundulus* maintained on a white background, melanin-concentrating material presumably would be secreted continuously by the nerve terminals and would accumulate in the surrounding tissues during the period of long-term adaptation. When such a fish was transferred to a black container, the melanin-dispersing principle would not only have to excite dispersion but would also have to compete with the pigment-concentrating material already accumulated in the tissues. The operation of the darkening substance would, therefore, be hampered and slower than normal. If such a fish were then returned to a white container, Parker and Brower hypothesized, blanching would be faster than normal because the stored concentrator would work in favor of lightening. Similar logic was used to explain the results obtained with fish kept in black containers.

The color change system of the eel *Anguilla* represents a more complicated type than that of *Fundulus*. Whereas in *Fundulus* nerves alone appear normally to regulate the chromatophores, in *Anguilla* nerves and hormones are involved. The American eel *Anguilla rostrata* darkens in 90 min and lightens in a little over 3 hr (Parker, 1945). Parker (1943) diagrammed the major features of the color change system of *Anguilla* (Fig. 31). He assumed that the dorsal retina, DR, of the eye would be con-

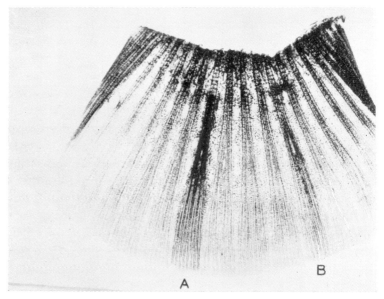

Fig. 30. Tail of a moderately pale killifish, *Fundulus heteroclitus*, showing a newly excited caudal band fully dark (A) and a re-excited caudal band moderately dark (B). The re-excited band when first cut was as dark as the newly excited one seen in this tail. It was then allowed to blanch fully after which it was re-cut. It then darkened considerably, but not completely. (From Parker, 1941.)

cerned with blanching, whereas the ventral retina, VR, would be concerned primarily with darkening. The eye of an eel, on a white background, W, would be stimulated partly by direct light, D, from the overhead source, L, and partly from scattered light, S, reflected from the background. The direct light would fall on the ventral retina and the scattered light from the white background on the dorsal retina, resulting in a pale eel. In contrast, the eye of an eel on a black background, (*B*), receives practically all of its stimulation directly from the light source because the black background absorbs so much and reflects so little, about 1/60th, of the incident illumination. Under such a circumstance, the fish darkens. At least four reflex tracts must be postulated to explain how stimulation of the eye results in a change in the distribution of pigment within the melanophores. The first arc, 1, runs from the ventral retina through the brain to the meta-adenohypophysis of the pituitary gland, *P*, where MSH is presumably released. MSH is then carried through the blood and lymph, *B*, to the melanophores whose pigment is then caused to disperse (DM), thereby darkening the eel. This is called the retino-pituitary arc. The second arc, 2, called the retino-cholinergic arc, also begins in the ventral retina, proceeds through various central nervous organs, CNO, and as cholinergic fibers, *C*, of the autonomic nervous system innervates the melanophores. Stimulation of these fibers also results in melanin dispersion. A third arc, 3, the retino-adrenergic one, runs from the dorsal retina through the central nervous organs to innervate the melanophores with adrenergic fibers, *A*, stimulation of which results in concentration of melanin, CM, thus a pale fish. The final arc, 4, proceeds from the dorsal retina to the brain and then to the part of the pituitary where a melanin-concentrating principle is released. This arc has been called the retino-tuberal arc, because the pars tuberalis (Waring, 1940) is thought to be the source of this material. The pars tuberalis of fishes according to the terminology of Pickford and Atz (1957) is the pro-adenohypophysis. The evidence of Waring (1940) for a lightening principle is based on essentially the same kind of experiments as those used to demonstrate a melanin-concentrating principle in elasmobranchs (Waring, 1938). Hypophysectomized eels are much more sensitive to MSH than intact specimens.

Another fish whose color change system has been examined in detailed fashion is the catfish *Ameiurus. Ameiurus nebulous* contains one type of pigment cells, melanophores (Parker, 1934b). The rates of melanin dispersion and concentration are very slow.

Migration of this pigment from the concentrated to the dispersed condition requires 15–24 hr; the reverse process requires 24–36 hr. Denervation of portions of the caudal fin results in the formation of black caudal bands some of which persist for 7 days. This response Parker believed was due to stimulation of dispersing nerve fibers. Olive oil extracts of skin from *Ameiurus* contain a substance that causes melanin dispersion (Parker, 1935c). Presumably, the appearance of a caudal band was due to release of this substance by the cut pigment-dispersing nerves. Evidence was also adduced by Parker (1934b) for a pigment-concentrating fiber innervating the melanophores. By appropriate electrical stimulation, Parker and Rosenbleuth (1941) were able to excite differentially the pigment-concentrating and pigment-dispersing fibers.

MSH also appears to be normally involved in the color changes of *Ameiurus*. In 1940 Parker reported that stimulation by light of skin photoreceptors in *Ameiurus* results in release of intermedin. However, in the normal darkening process, the dispersing nerves appear to initiate pigment migration; their action is then followed and substantially supplemented by intermedin (Parker, 1941). Abramowitz (1936) had previously reported that melanin dispersion is not as great in hypophysectomized *Ameiurus* on a black background as in intact specimens in similar surroundings.

As he did for the eel *Anguilla*, Parker (1943) diagrammed the interrelationships of the several components of the more complicated color change system in the catfish *Ameiurus* (Fig. 32). The first, second, and third arcs described for the eel occur in *Ameiurus*, but no evidence for the fourth arc has been obtained. Illumination of the integument, I, will cause melanin dispersion, as occurs in blinded specimens. As mentioned above, this response is not due to direct stimulation of the melanophores, but rather begins in integumentary photoreceptors, passes to the central nervous organs, and then to the pituitary gland where MSH is released (Parker, 1940). The fifth, 5, also a darkening one, is called the dermo-intermediate excitatory arc. The excitatory effect of this arc presumably is inhibited when illuminated catfish are on a white background. Parker postulated the presence of an inhibitory arc, 6, in the catfish and called it the retino-intermediate inhibitory arc.

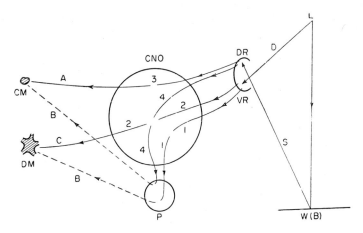

FIG. 31. Diagram of the melanophore system in the eel *Anguilla*. 1, retino-pituitary arc; 2, retino-cholinergic arc; 3, retino-adrenergic arc; 4, retino-tuberal arc. A, adrenergic fibers, B, blood and lymph; (B), black background; C, cholinergic fibers; CM, melanin concentration; CNO, central nervous organs; D, direct light; DM, melanin dispersion; DR, dorsal retina; L, light source; P, pituitary gland; S, beam of scattered, reflected light; VR, ventral retina; W, white background. (From Parker, 1943.)

The chromatophore system of the European minnow *Phoxinus laevis* appears to be very similar to that of *Anguilla*, some evidence having been obtained for pigment-concentrating and dispersing pituitary products as well as nerve fibers with corresponding functions. However, the evidence for a normal role of MSH is weak. Von Frisch (1911) first described the melanin-concentrating nerve fibers. These pass from the brain to the spinal cord and leave the latter at the level of the fifteenth vertebra to enter the autonomic chain. They then run forward and backward to supply the melanophores on the body surface. In the head a branch of the trigeminal nerve cooperates with the other fibers. These melanin-concentrating fibers are controlled by a center in the medulla oblongata which in turn receives inhibitory stimulation from a center in the brain. Later, Giersberg (1931) found evidence for melanin-dispersing fibers; hence these melanophores are dineuronic as in *Fundulus*. Von Gelei (1942) showed that the melanin-dispersing fibers left the spinal cord with the first or second spinal nerve. Gray (1956) also favors the view that the melanophores of *Phoxinus* are dineuronic. He studied the formation of caudal bands and found that hypophysectomy did not interfere with darkening of the tail band following nerve section, just as in *Fundulus*. Healey (1951, 1954) found that cutting the spinal cord between the fifth and twelfth vertebrae caused an immediate darkening. This result was interpreted as interruption of the melanin-concentrating nerve tracts rather than stimulation of melanin-dispersing fibers. After some time physiological color changes resumed, but they were slower than in unoperated specimens and presumably were strictly hormonal. In 1940 he obtained some evidence for a pigment-concentrating principle. For example, crude pituitary extracts and implants of entire pituitary produced a transitory pallor. His evidence (Healey 1951) for a darkening hormone was based on analysis of the rates of melanin dispersion and concentration under several different experimental conditions.

Abolin (1925) had found that epinephrine caused melanin concentration in *Phoxinus*. Reidinger (1952) reported that this drug would indeed concentrate pigment in melanophores of *Phoxinus*, but not in the xanthophores. Pituitary extracts on the other hand, dispersed the xanthophore pigment. She felt that with respect to hormones, adaptation to a yellow background was due to melanin concentration by epinephrine and yellow pigment dispersion by MSH. Giersberg (1931) had reported that these xanthophores are not innervated and concluded they must be under hormonal control. Köhler (1952) found that adrenocorticotropic hormone dispersed the pigment in the melanophores of intact adrenalectomized and hypophysectomized specimens of *Phoxinus* as well as in isolated pieces of skin. Through injection of ACTH he was able to evoke the nuptual coloration which depends upon activation of melanophores xanthophores, and erythrophores. Pickford (1956), on the other hand, found no response to ACTH in hypophysectomized *Fundulus heteroclitus.*

As far as other fishes are concerned, Enami (1939) reported briefly on the pigmentary reactions of a marine catfish, *Plotosus anguillaris.* This fish showed a very poor background response, very little lightening occurred when placed on a white background. Hypophysectomy also had little or no effect on the degree of melanin dispersion. However, epinephrine caused a generalized blanching. Electric shocks also caused blanching, presumably through stimulation of pigment-concentrating nerve fibers. In 1940 Enami reported the effects of epinephrine on several other fishes. A blanching reaction was observed among specimens of *Anguilla japonica, Aplocheilus latipes, Chaenogobius macrognathus, Chasmichthys dolichognathus, Dictyosoma burgeri, Gobius similis, Misgurnus anguillicaudatus,* and *Salaris enosimae.* Unexpectedly, however, the catfish *Parasilurus asotus* darkened in response to epinephrine. Such a response is highly unusual especially since pigment-concentrating nerve fibers in teleosts are supposed to be adrenergic (Parker, 1943). A similar darkening reaction was noted by Breder and Rasquin (1950, 1955) when they injected epinephrine into an angel fish, *Chaetodipterus faber;* the pigment in the dermal melanophores dispersed but the melanin in the iris and meninges behaved as expected and concentrated. In 1955 Enami reported that norepinephrine had the same effect as epinephrine on *Parasilurus.* He also stated that acetyl choline blanched this fish, another unexpected reaction, since melanin-dispersing fibers are supposed to be cholinergic (Parker, 1943).

Müller (1953) found that thyroxin caused melanin dispersion in the goldfish, *Carassius auratus.* Briseno-Castrejon and Stevens Flores (1955) reported that pituitary extracts of *Carassius* caused melanin dispersion in this fish.

Weisel (1950) tested the effects of pituitary extracts from four fishes (the sheepshead, *Pimelometopon pulchrum;* the barracuda, *Sphyraena argentea;* the yellowfin tuna, *Neothunnus macropterus;*

and the white sea bass, *Cynoscion nobilis*) on the melanophores of a variety of other fishes. The extracts darkened the round stingray, *Urobatis halleri*, and the black bullhead, *Ameiurus melas*, and lightened the mosquito fish, *Gambusia affinis*, the California killifish, *Fundulus parvipinnis*, the opaleye, *Girella nigricans*, the mudsucker, *Gillichthys mirabilis*, the green sunfish, *Lepomis cyanellus*, and the grunion, *Leuresthes tenuis*. A possible explanation of this difference in response to pituitary extracts among fishes is that the fish pituitary contains a pigment-concentrating substance and a pigment-dispersing principle and that some fishes respond better to the darkening substance and other fishes better to the lightening material. Weisel found that beef pituitaries and commercial pituitrin (Parke-Davis) had no effect on the melanophores of those fishes in which fish pituitaries had induced melanin concentration. These fish pituitaries, in addition to their effects on the melanophores, caused dispersion of the xanthophore pigment in *Gambusia*, *Fundulus*, and *Girella*.

Robertson (1951a) determined the responses of the melanophores in the rainbow trout, *Salmo gairdneri*, to a variety of physical and chemical factors. He found that melanin concentration was induced by asphyxia, bright light, high temperature, potassium chloride, epinephrine, and mammalian thyroid. Dispersion was produced by acetyl choline and extracts of the posterior lobe of mammalian pituitary glands. Acetyl choline produced a generalized response whereas the pituitary extract produced a response only in the region where it was injected into the trout. Robertson (1951b) also noted that pressure would induce paling of a rainbow trout. He postulated this response was due to liberation of potassium by muscles under pressure and that the liberated potassium was responsible for the lightening reaction.

Kinosita (1953) believes that migration of granules in melanophores of the medaka, *Oryzias latipes*, is the result of a potential difference between the center of the cell and the peripheral branches (Fig. 33). Intracellular electrophoresis revealed that the melanin granules are negatively charged. Presumably, a melanin-dispersing agent would cause the center of the cell to become negatively charged. Reversal of the polarity would cause the melanin to concentrate.

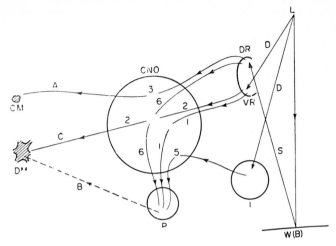

FIG. 32. Diagram of the melanophore system in the catfish *Ameiurus*. Symbols same as in Fig. 30. In addition, 5, dermo-intermediate excitatory arc; 6, retino-intermediate inhibitory arc; I, integument. (From Parker, 1943.)

Enami (1955) found two antagonistic responses when aqueous extracts of the hypothalamus and the pituitary gland were injected into a specimen of *Parasilurus asotus* whose melanin was in an intermediate degree of dispersion as a result of hypophysectomy. A marked localized pallor appeared at the site of injection while the rest of the body darkened. In a few instances, highly concentrated extracts produced a generalized blanching. He proposed that this material be called the melanin-concentrating hormone (MCH). The meso-adenohypophysis contained a higher concentration of MCH than any other part of the pituitary gland. Cutting the pituitary stalk resulted in a decrease in the MCH content of the meso-adenohypophysis. Enami, therefore, concluded that MCH is a product of neurosecretory cells in the hypothalamus and is transported via axons for storage in the meso-adenohypophysis. The portion of the hypothalamus that contained the highest concentration of MCH was the nucleus lateralis tuberis.

The melanophores and erythrophores of the bitterling, *Rhodeus amarus*, are innervated by at least a pigment-concentrating fiber. Their pigments disperse after section of the spinal cord, a result similar to that described above for *Phoxinus*. Physiological color changes in the adult males lead to nuptual coloration, males show very little ability to adapt to different backgrounds (Brantner, 1956). In contrast, females show excellent background responses. Extracts of the posterior lobe of the pituitary gland caused maximal dispersion of melanin and an increase in the number of melanophores. Testis extracts and extracts of anterior and posterior lobes of the pituitary caused an increase in the number of erythrophores in males without affecting their state of dispersion. Epinephrine caused a decrease in the number of erythrophores.

Umrath and Walcher (1951) reported that the melanophores and lipophores of *Macropodus opercularis* are not responsive to pituitary extracts even after spinal section. Apparently in this fish color changes depend only on nerves. Umrath (1957) found that physiological color changes of the bitterling, *Rhodeus amarus*, could be induced by injection of acetyl choline (darkening), epinephrine (lightening), and norepinephrine (lightening). Intermedin caused an increase in the number of melanophores without dispersing the melanin, further evidence for the independence of physiological and morphological color changes. In 1959 Umrath reported that ACTH evoked the nuptual coloration in *Phoxinus laevis*, *Rhodeus amarus*, and *Macropodus opercularis*. ACTH was least effective on *Macropodus*. The response of *Phoxinus* confirmed the earlier finding of Köhler (1952). Parasympatheticomimetic drugs caused darkening of *Rhodeus* and *Phoxinus*. Umrath (1959) postulated that intermedin, epinephrine, ACTH, and the sex hormones normally cooperate to produce the nuptual coloration.

Rasquin (1958) determined the responses of 35 species of teleosts injected with epinephrine and intermedin. These fishes could be grouped in three categories according to the reaction of the melanophores to epinephrine: (1), fishes in which concentration of pigment occurred in all the melanophores, (2), fishes in which only the internal melanophores showed pigment concentration, and (3), fishes in which none of the melanophores responded. In none of these fishes did melanin dispersion occur. Dispersion of melanin in response to intermedin was observed in only 4 species, *Ameiurus nebulosus, Gambusia* sp., *Astyanax mexicanus*, and *Atherina stipes*. The remaining 31 species showed no melanophore reaction to intermedin. Leucophores responded to epinephrine by dispersion of the guanine granules. The pigments in the xanthophores and erythrophores of every species dispersed in response to MSH. The responses of these lipophores to epinephrine, however, were inconsistent; in some fishes dispersion of pigments occurred, in others concentration.

Fujii (1960) determined the effect of atropine on the melanophores of the goby *Chasmichthys gulosus*. This drug accelerated the dispersion of melanin that had been concentrated by epinephrine. This effect was noted when intact or denervated melanophores were used, thereby suggesting to Fujii that the agent acts directly on the cells. The possibility remains, however, that the atropine acted on the nerve endings attached to the melanophores.

As discussed above, the mechanisms controlling the chromatophores of fishes are extremely diverse. Much of the available information concerns the melanophores alone. A good deal of research remains to be done before we have full comprehension of the control of the other chromatophore types in fishes. More information is needed in order to understand fully how the nerves and hormones cooperate to produce the final coloration of a fish. MSH appears to activate all chromatophore types.

CHROMATOPHORES OF AMPHIBIANS

Considerable information has accumulated concerning chromatophores of amphibians. Investigators in the nineteenth century held the opinion that amphibian color changes are under nervous control. However, endocrines are now known to be the main regulators of these chromatophores. In some species nerve fibers seem to exercise a slight control over the melanophores; in others none at all.

Color changes in amphibians are not very striking. Nevertheless they are readily detected. Changes are relatively slow, usually requiring a matter of hours, if not days. The vast majority of amphibians possess three types of chromatophores, xanthophores, guanophores, and melanophores. The latter are the major contributors to the color change process. They occur in the epidermis and dermis of tadpoles and adults. Amphibian melanophores are usually highly branched but epidermal melanophores may be filamentous, having few branches (Herrick, 1933). Recently, Volpe and his associates described the distribution of melanophores in larvae of three amphibians. A dense concentration of filamentous melanophores occurs in the epidermis of *Rana palmipes* from Costa Rica, thereby rendering the tail fins cloudy even when the pigment in the dermal melanophores is not dispersed (Volpe and Harvey, 1958). Color changes of these tadpoles, however, depend primarily upon the degree of pigment dispersion in the dendritic dermal melanophores which occur singly or in clusters, the latter condition leading to the gross appearance of "spots" on the tail. These clusters of melanophores first appear at stage II (an early limb bud stage) of the system of Taylor and Kollros (1946) for staging larvae of amphibians. The dorsal portion of the tail musculature of the larval oak toad, *Bufo quercicus*, is covered with melanophores; the greater part of the ventral half is devoid of them (Volpe and Dobie, 1959). The tail fins and ventral half of the tail musculature contain very few melanophores. The greatest concentration of xanthophores and leucophores occurs along the dorsal edge of the tail musculature and at the central border of the melanophore-darkened dorsal musculature. Prior to resorption of the gills the dorsal tail musculature of the bird-voiced, tree frog, *Hyla avivoca*, becomes conspicuously marked by a series of "saddles" consisting of an immobile pigment that appears red by reflected illumination (Volpe, Wilkens, and Dobie, 1961). Xanthophores then develop distal to the "saddles" and may completely mask the underlying pigment when the yellow pigment is maximally dispersed. The tail musculature has an irregular pattern of melanophores. In early development, the dorsal half is darkened by melanophores and the ventral half contains scattered melanophores. Continual differentiation of melanophores results in a progressive narrowing of the unpigmented ventral portion of the tail musculature. The transparent tail fins are flecked with melanophores and xanthophores.

Several investigators in the twenties and thirties showed that the intermediate lobe of the pituitary contains a melanin-dispersing principle. The results of Hogben and his coworkers may be used to exemplify these findings. Hogben and Winton (1923) used the frogs *Rana temporaria* and *Rana esculenta*. Hypophysectomized frogs remained permanently pale as a result of maximum concentration of melanin. Hogben and Winton found no evidence to support the possibility that nerves play a role in controlling amphibian melanophores. Neither section nor stimulation of peripheral nerve trunks was followed by local pigmentary changes. Hogben and Slome (1931) found that hypophysectomized *Xenopus laevis* had a lower threshold to pituitary extracts and the darkening produced was much greater than in intact animals (Fig. 34). However, the response to pituitary extracts of *Xenopus* from which only the posterior lobe had been removed was less than the response of intact specimens. As an explanation of these observations they postulated the existence of a melanin-concentrating principle in the pars tuberalis of intact specimens. This substance would then be available to antagonize injected darkning material. Several experiments with *Xenopus*, as with *Rana*, revealed no need to assume that the nervous system plays any part in the control of color changes. Among the experiments performed were (1), section of the entire nerve supply to a hind limb and (2), spinal section at several levels. Some support for the hypothesis that a lightening substance is elaborated by the pars tuberalis was obtained by Steggerda and Soderwall (1939). These investigators found that when the pars tuberalis of *Rana pipiens* is destroyed by local cautery, the frog is unable to blanch completely. Parker (1948), however, questioned the validity of their findings. He felt that the cautery may have excited the intermediate lobe sufficiently to cause a low level release of intermedin, leading to a slight darkening. Bors and Ralston (1951) found that adult *Xenopus laevis* lightened when injected with pig or human pineal gland extracts. Jørgensen and Larsen (1960) wished to test the hypothesis of Hogben and Slome (1931) that the pars tuberalis of amphibians secretes a melanin-concentrating principle. The neuro-intermediate lobe of *Bufo bufo* and *Xenopus laevis* was denervated by cutting the wall of the third ventricle which separates the median eminence from the infundibulum. The pars tuberalis was left intact, whereas nervous regulation of intermedin secretion was interrupted. If lightening hormone were present

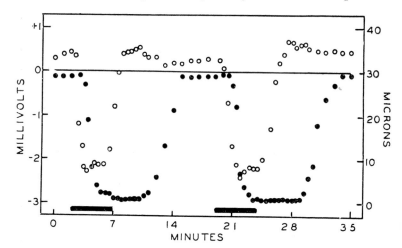

FIG. 33. Results of simultaneous determinations of change in melanophore potential (circles) and the length of a process of the same melanophore (dots) when the medium was altered from physiological saline to isotonic KCl solution (solid rectangles). Both microelectrodes were in the melanophore, one tip in the center and the other in the branch. Distance between electrodes: 35μ. The potential differences are for the center relative to the peripheral branch. (Redrawn from Kinosita, 1953.)

in the pars tuberalis, then a change from a black to a white background should result in lightening. Seven of eight *Xenopus* and 20 of 21 *Bufo* showed no such background response. Paling, they concluded, must be due normally to disappearance of intermedin from the blood and not to a lightening substance from the pars tuberalis. They feel that secretion of intermedin is regulated by both inhibitory and excitatory nerves.

Parker and Scatterty (1937) described the control of color changes in the common leopard frog, *Rana pipiens*, the American counterpart of the European *Rana temporaria*. Darkening in *Rana pipiens* is induced by intermedin. Frogs become maximally pale when injected with epinephrine or when hypophysectomized. Intact animals are lighter when on a white background than on a black background, but rarely is the melanin maximally concentrated or maximally dispersed. Blood serum from dark frogs darkens pale ones, but serum from pale frogs does not lighten dark ones. It would appear that the normal control of melanin migration is due to intermedin alone. However, Kropp (1927) has shown that nerves can play a small, but significant, role in the control of melanophores in *Rana pipiens*. Upon severing nerves in a leg, he noted a slight dispersion of melanin usually occurred in the web of the foot. Stimulation of spinal nerves and sympathetic roots in the sciatic plexus led to paling of the portion of the skin supplied by the stimulated nerve fibers.

Rowlands (1950) found that dampness induced melanin dispersion in the frog *Rana temporaria* maintained on a white background and that dehydration caused paling even in specimens on a black background. The greater the degree of immersion of a frog, the greater the extent of pigment dispersion throughout the skin. Presumably, skin receptors are involved in the response. However, male frogs in the breeding season were an exception to this rule; they did not exhibit this response to moisture. In 1952 Rowlands, reported that destruction of the eyes, severing the optic nerves, or cauterization of the optic chiasma had no effect on the moisture response. Parker and Scatterty (1937) had found that *Rana pipiens* showed no response to humidity. In 1954 Rowlands reported that the toad *Bufo bufo* also shows no response to humidity. She also noted that melanin of hypophysectomized toads remains permanently concentrated.

It has been known for several years that skin extracts of amphibians would cause melanin concentration. For example, Strohm

(1936) showed this was the case with *Rana esculenta*. More recently, Sieglitz (1951) studied this phenomenon in *Rana temporaria*. She showed by chemical and biological techniques that this melanin-concentrating substance is neither epinephrine nor norepinephrine. The amount of material that could be extracted was inversely proportional to the degree of melanin dispersion, i.e. the lighter the skin, the more that could be extracted. She believes this material is released from sympathetic nerve endings in the skin. Lightening of a frog, she felt, can be caused by removal of intermedin from the blood or by secretion of the lightening substance which can be used to antagonize the circulating intermedin.

Stoppani (1941, 1942) found that the peripheral nervous system played a secondary but nevertheless significant role in controlling color changes of the toad *Bufo arenarum*. Stimulation of peripheral nerve fibers caused melanin concentration and dispersion of the pigment in guanophores. In 1954 Stoppani, Pieroni, and Murray using the same species confirmed and extended these earlier observations. This action of the peripheral nerves is opposite to that observed when pituitary extracts are administered. The responses of subcutaneous melanophores were sluggish in comparison with those of the intracutaneous ones. The action of the peripheral nerves seemed to be mediated by an epinephrine-like substance; norepinephrine and the cutaneous secretion which contains epinephrine, among other things (Deulofeu, 1935), had no apparent physiological significance. Paling as a result of nerve stimulation was not as rapid as that which occurs after hypophysectomy.

Burgers, Boschman, and Van de Kamer (1953) and Burgers (1956) reported the results of a detailed study of the melanophores in the South African clawed toad, *Xenopus laevis*. As mentioned above, Hogben and Slome (1931) had already shown the presence of a melanin-dispersing substance in the pituitary of this amphibian. Burgers and his co-workers found that the melanophores of *Xenopus* differed from those of most other amphibians that had been investigated. Injection of epinephrine caused melanin dispersion whereas the typical amphibian response to this substance is blanching. In addition, excitement stimuli resulted in darkening rather than pallor. Amphibians usually respond to excitement stimuli by blanching. The only other exception to this rule was reported by Siedlecki in 1909 for the Javanese "flying frog" *Polypedatus reinwarti*. Ketterer and Remilton (1954b) also noted this excitement darkening reaction in *Xenopus*. Skin glands of *Xenopus* secrete a substance that causes darkening of the skin (Burgers, 1956; Burgers and Van Oordt, 1956). The melanin-dispersing principle in the skin secretion of *Xenopus laevis* has been identified by chromatographic techniques as serotonin (5-HT) by Van de Veerdonk (1960) and Van de Veerdonk, Huismans, and Addink (1961). Erspamer (1954) previously had found large amounts of this substance in amphibian skin. Excitement darkening was not due to intermedin but probably to serotonin and epinephrine. Burgers (1956) also noted that adrenocorticotropic hormone caused melanin dispersion in *Xenopus*.

Chang (1957) reported that tadpoles and adults of *Xenopus laevis* blanched when injected with thyroxine. This effect, however, did not appear to be a direct response of the melanophores to thyroxine, but rather an indirect one involving a cholinergic substance which in turn appeared to control release of intermedin from the pituitary. Evidence in favor of the mediation of a cholinergic principle is the fact that the response to thyroxine was inhibited if test animals were injected with atropine, which is anticholinergic, 1 hr prior to injection of the thyroxine. Injection of acetyl choline without thyroxine also resulted in blanching. Triidothyronine had no effect on the melanophores.

It had been known for a long time (Allen, 1916) that hypophysectomy of *Rana pipiens* tadpoles resulted in permanently pale individuals. More recently, however, Bagnara (1957) found that this is not true of the melanophores in the ventral fin of the tail of tadpoles of *Xenopus laevis*. When normal or hypophysectomized tadpoles were placed in a darkroom for 30 min the pigment in the tail melanophores dispersed, rendering the tail dark (Fig. 35). When these tadpoles were again illuminated, the melanin concen-

9

trated to the original condition in 5–6 min. This behavior was a direct response of the melanophores to illumination, the eyes were not involved. Van der Lek, De Heer, Burgers, and Van Oordt (1958) found that the degree of melanin concentration in

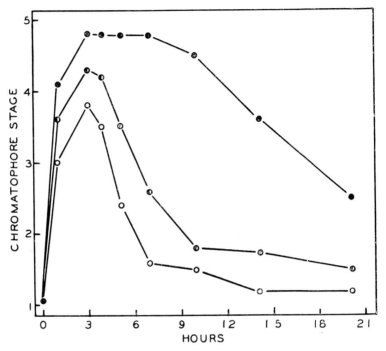

FIG. 34. Responses of *Xenopus laevis* to MSH. Circles, posterior lobe of pituitary removed; half-filled circles, intact specimens; dots, hypophysectomized specimens. (Redrawn from Hogben and Slome, 1936.)

the tails of *Xenopus* larvae was a direct function of intensity of incident illumination. Bagnara (1960a) noted that these melanophores showed their characteristic response to illumination almost from their first appearance in the tail; however, the response was not as striking as in older tadpoles.

In contrast to the ventral fin melanophores, the pigment in the integumental and deep melanophores of the body concentrated when intact or blinded tadpoles were placed in darkness. This response of the body melanophores was first reported by Bles (1905). Bagnara (1960a) found that the response of the body melanophores, although obvious macroscopically, was not as striking as the response of the tail melanophores. Further research by Bagnara (1960b) revealed that the pineal gland is involved in control of the response of the body melanophores when tadpoles are placed in darkness (Fig. 36). Pinealectomy abolished the response of the body melanophores but the response was induced

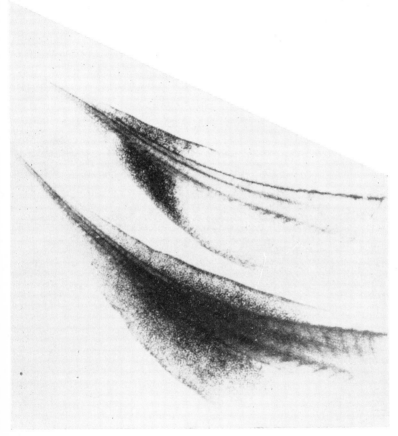

FIG. 35. Tails of intact larvae of *Xenopus laevis*. Top, tail-darkening reaction at tip but not in illuminated center of tail; bottom, height of tail-darkening reaction. (From Bagnara, 1957.)

by melatonin at a concentration of 0.01 mg/ml aquarium water. Presumably, when tadpoles of *Xenopus* were placed in darkness the pineal gland released melatonin which in turn caused concentration of the pigment in the head melanophores. Upon return to light, the pineal stopped secreting this principle. Bagnara (1961) noted that the body-lightening reaction first appeared at larval stage 35 of the system of Nieuwkoop and Faber (1956). It would be interesting to learn why the tail melanophores do not respond to melatonin from the pineal gland when the body-lightening reaction is proceeding in darkness.

Wright (1954b, 1955) studied with photoelectric recording equipment the behavior of the melanophores in excised portions of skin from the frogs *Rana pipiens* and *Rana clamitans*. The melanin dispersed in response to intermedin and concentrated in the presence of epinephrine. The rates of melanin concentration with epinephrine and dispersion with MSH increased with increase in temperature. Sodium iodoacetate, malonate, and fluoride at least partially inhibited melanin concentration. Triphenyltetrazolium chloride inhibited the responsiveness of the melanophores to intermedin; energy exchange may have been blocked by this reagent. Wright postulated that the energy of glycolysis is necessary for blanching. However, Lerner and Takahashi (1956) believe that pigment dispersion is the important energy consuming part of the process. They claim ATP causes melanin dispersion in frogs. Presumably, MSH "activates" ATP. Shizume, Lerner, and Fitzpatrick (1954) found that epinephrine and norepinephrine inhibited the action of MSH on skin of *Rana pipiens*. Cortisone and hydrocortisone did not interfere.

In 1957 Kahr and Fischer reported that lysergic acid diethylamide (LSD) and 5-oxytryptamine (5-OT) produced a darkening of intact frogs, *Rana temporaria*, but had no effect on hypophysectomized individuals. These investigators concluded, therefore, that the action of these drugs was indirect via the pituitary rather than directly on the melanophores. They postulated that these drugs interfered with release of ACTH since this hormone has a melanin-dispersing action. In view of the fact that drugs such as epinephrine act oppositely on *Xenopus* as compared with their action on most other amphibians Burgers, Leemreis, Dominiczak, and Van Oordt (1958) thought it would be interesting to determine the action of LSD on *Xenopus*. They found that LSD caused a marked pigment concentration in the melanophores of *Xenopus*. They also attributed this response to an indirect action, but postulated that it probably acted by inhibiting the release of intermedin from the pituitary gland. It would be interesting to determine how LSD and 5-OT really act.

Several investigators have studied color changes in tree frogs. Parker (1930) found that epinephrine blanched *Hyla versicolor* and commercial pituitrin darkened it. Müssbichler and Umrath (1950) who studied the chromatophores of *Hyla arborea* found that its color changes are determined primarily by hormones, secondarily by nerves. A preponderance of epinephrine in the blood caused the skin to become light, in extreme cases nearly yellow. High concentrations of intermedin caused the frog to become dark green, in extreme cases nearly black. Gray body color appeared to be determined primarily by high concentrations of epinephrine and intermedin. Electrical stimulation of the skin covering intact and denervated limbs caused dispersion of yellow pigment and melanin concentration. This response was thought to be due to stimulation of autonomic nerves in the skin. Sulman (1952a, b) noted that ACTH caused *Hyla arborea* to change from a bright green to a dark brown. *Hyla cinerea* was shown by Edgren (1954b) and Patton and Teague (1959) to respond to intermedin by darkening.

Novales (1959) reported some very interesting data on the action of purified porcine β-MSH. He found that sodium was required for MSH to act on the melanophores of *Rana pipiens* and *Rana catesbiana* (Fig. 37). Also, pure α-corticotropin did not darken frog skin in the absence of sodium. The requirement for sodium was specific, neither potassium, choline, lithium, nor guanidinium was able to replace sodium in permitting a response to MSH. Sodium was not required for melanin concentration to occur.

Novales interpreted these findings in the following way. MSH increases the permeability of the melanophores to sodium which causes an increase in the internal sodium concentration in proportion to the external MSH concentration. Melanin dispersion is then triggered by the entry of sodium, the degree of dispersion being proportional to the amount of sodium that has entered.

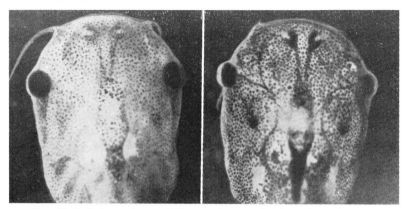

Fig. 36. Left, intact larva of *Xenopus laevis* at height of body-lightening reaction after one hour in darkness. Right, pinealectomized larva after one hour in darkness. (From Bagnara, 1960.)

Novales, Novales, and Zinner (1960) found that sodium could not be replaced by ammonium and tetraethylammonium Ringer solutions. Novales (1960) found that tissue-cultured embryonic melanophores of the California newt, *Taricha torosa*, would respond to epinephrine by pigment concentration and to MSH by formation of processes and peripheral migration of the pigment. The responses were the same as are observed *in vivo*. Melanophores of *Ambystoma maculatum* and *A. tigrinum* in tissue culture respond to MSH by melanin dispersion and expansion in surface area (Novales and Novales, 1961). There appeared to be a sodium requirement for the action of MSH on these chromatophores also. Replacement of sodium by potassium in the presence of MSH was followed by melanin concentration. Kulemann (1960) has also studied the behavior of embryonic melanophores in tissue culture. She used *Xenopus laevis* and found that, as *in vivo*, MSH, ACTH, and epinephrine caused melanin dispersion (Fig. 38). Strong light also caused melanin dispersion.

Lerner and Takahashi (1956) favored a theory that arose from the high pressure studies of Marsland (1944), namely that MSH causes a change of a melanin-bearing cytoplasmic gel to a sol.

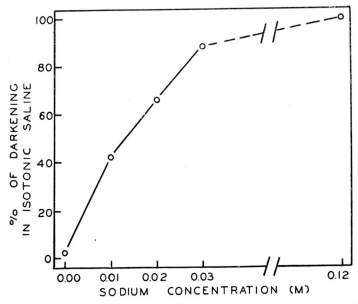

FIG. 37. Relationship between sodium concentration and the response of paled frog skin to MSH. (Redrawn from Novales, 1959.)

How MSH could accomplish this change is unknown, but Wright and Lerner (1960a) feel that MSH directly or indirectly produces a change in the specific ionic permeability of the melanophore membrane.

Davey (1959) reported that serotonin has a slight but distinct melanin-dispersing action on *Rana pipiens*. LSD caused lightening. In 1960 Davey set forth a novel theory concerning the mode of action of MSH on melanophores of *Rana pipiens*. Davey feels that intermedin has an indirect rather than a direct action on melanophores. He postulated that intermedin acts on elements in the skin to cause the release of a bound indolalkyamine such as serotonin, which in the free state would cause melanin dispersion. In the absence of intermedin, no indolalkylamine would be liberated; any indolalkylamine free in the skin would be metabolized and consequently the melanin would concentrate. Davey was able to potentiate the action of MSH by adding iproniazid, an inhibitor of monoamine oxidase, and phenylthiourea, an inhibitor of the phenol oxidases. Both of these enzymes could be involved in the destruction of indolalkylamines. However, Lerner, Case, Takahashi, Lee, and Mori (1958) and Wright and Lerner (1960b) stated that serotonin antagonized the action of MSH. This conflict concerning the action of serotonin certainly needs clarification. Wright and Lerner (1960b) also noted that melatonin, norepinephrine, hydrocortisone, acetyl choline, thyroxine, and triiodotyrosine can reverse the darkening action of MSH on *Rana pipiens*.

Some studies on long-term adaptation have been performed with *Rana pipiens*. Kleinholz and Rahn (1940) found no difference in the intermedin content of pituitaries from frogs kept in light and in darkness. Frieden and Bozer (1951) found that administration of hog intermedin resulted after 4 weeks in a 20% decrease in the melanin content of the skin. Subsequently, there was a rapid increase reaching a level 40% above normal after a total of 8 weeks. The initial drop was surprising. The increase might have been due to either the high concentration of intermedin in the tissues inducing melanin synthesis or maximal dispersion of the melanin inducing synthesis of additional pigment in the melanophores. Ortman (1954) found an inverse relationship between the amounts of periodic acid Schiff (PAS)-positive material and acid hematein positive material in the pars intermedia. When frogs were maintained on a black background, the PAS-positive material in the pars intermedia was depleted and the acid hematein positive material increased. The opposite situation was observed in the pars intermedia of specimens adapted to a white background. However, Ortman (1954, 1956) reported that the amount of intermedin in the pars intermedia did not change when specimens were kept in black and in white containers for periods up to 528 hr. Injections of intermedin also had no effect on the amount of stored hormone (Ortman, 1956).

CHROMATOPHORES OF REPTILES

Control mechanisms for reptilian color changes show almost as much variation as occurs in fishes. The most dramatic color changes among reptiles are exhibited by lizards. Most of the information concerning reptilian chromatophores centers around the lizards *Chamaeleo*, *Phrynosoma*, and *Anolis*. Very little research on reptilian chromatophores has been done recently.

As long ago as 1852 Brücke postulated that color changes in *Chamaeleo* are controlled directly by nerves. Color displays of chameleons are the most complex among the reptiles. Zoond and Eyre (1934) stated that the South African chameleon, *Chamaeleo pumilis*, has an integumentary pattern of stripes, patches, and tubercles of different colors within each of which pigment concentration or dispersion may occur independently of other regions of the body. These investigators described in detail the various colors each part of the pattern may assume. Hogben and Mirvish (1928) learned that *Chamaeleo pumilis* exhibits excitement pallor. They also observed generalized blanching following faradic stimulation of the roof of the mouth or cloaca. After the spinal cord was cut at any level in the trunk region, stimulation of the cloaca caused lightening only posterior to the cut. However, stimulation of the roof of the mouth after the cord was sectioned at any level anterior to the tenth or eleventh vertebra resulted in blanching anterior to the point of section only. However, stimulation of the roof of the mouth in animals whose cords had been sectioned posterior to the eleventh or twelfth vertebra resulted in a generalized paling of the entire body with the exception of a short portion at the tip of the tail. These results were interpreted in terms of pigment-concentrating nerves. Epinephrine also caused lightening. However, Hogben and Mirvish felt that epinephrine did not play a role in excitement pallor. Severing a nerve to an area of the skin resulted in that part of the skin becoming dark (Zoond and Eyre, 1934). This darkening can be interpreted as a release from inhibition by pigment-concentrating fibers or stimulation of pigment-dispersing fibers. However, there ts no direct evidence for pigment-dispersing fibers nor has the possibility that the pituitary is involved been adequately explored. The color change of the lizard *Phrynosoma* is not as striking as in *Chamaeleo*. The dorsal surface of *Phrynosoma* is marked by fiour or five transverse, dark bands set against a background that can be pale or dark (Fig. 39). The changes can best be observed

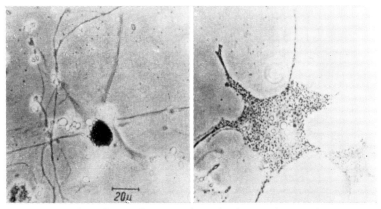

FIG. 38. Melanophores of *Xenopus laevis* in tissue culture. Left, concentrated melanin; right, after exposure to MSH. (From Kulemann, 1960.)

in the marginal dentate scales (Fig. 40). Melanin concentration in *Phrynosoma cornutum* and *P. blainvilli* occurs through the action of epinephrine and concentrating nerve fibers (Redfield, 1918; Parker, 1938b). No evidence could be obtained for pigment-dispersing fibers. Denervation of a portion of the skin did no result in darkening but injection of MSH did. Hypophysectomized lizards blanched completely in 17 hr.

Anolis carolinensis shows a color display ranging from dark brown to bright green. The intermediate colors are light brown yellow, yellowish green, and emerald green (Kleinholz, 1938a, c). Hypophysectomy results in permanent pallor; MSH darkens hypophysectomized individuals.

Analysis of the control mechanisms revealed that denervated regions of the skin underwent normal color changes. Electrical

FIG. 39. Two specimens of *Phrynosoma* showing extremes of blanching and darkening. (From Parker, 1938b.)

stimulation of intact specimens by placing one electrode in the mouth and the other in the cloaca resulted in a generalized darkening. However, when this experiment was performed with an hypophysectomized individual generalized darkening did not occur, but instead only the postorbital regions darkened and the body became mottled. This mottling consisted of clusters of black spots in an irregular pattern against the green color of the skin. Generalized darkening must depend, therefore, upon the pituitary and not upon innervation of the melanophores. Darkening of the postorbital regions was not under nervous control, having occurred even when this portion of the skin was denervated. Injection of epinephrine mimicked the mottled condition produced by electrical stimulation of hypophysectomized individuals. Additional evidence that epinephrine causes mottling was obtained when Kleinholz (1938a) found that hypophysectomized, adrenalectomized specimens did not show the mottling response. The activity of the melanophores in *Anolis* appears, therefore, to be regulated exclusively by hormones.

Kleinholz (1938c) found that epinephrine is not normally necessary for blanching. Rather, paling appears to be due to removal of MSH from the blood. The chromatic responses of adrenalectomized specimens do not differ from those of intact specimens. Rahn (1956) reported that increase in body temperature or decrease in the partial pressure of oxygen intensifies the response to injected epinephrine. Novales (1959) reported that darkening of *Anolis* skin by MSH is significantly reduced in the absence of sodium. Horowitz (1957) suggested that MSH causes dispersion of melanin in *Anolis* by producing an oxidation of sulfhydryl groups in the melanophore. The effect of MSH is decreased by pretreatment of melanophores with glutathione or cysteine (Fig. 41). Furthermore, sulfhydryl-binding agents such as mersalyl are able to produce pigment dispersion in melanophores of *Anolis*. Thus MSH is regarded as acting in a manner similar to that of mersalyl. Horowitz (1958), continuing his study of the melanophores in *Anolis*, noted that adenosine monophosphate, 2, 4-dinitrophenol, and other uncouplers of oxidative phosphorylation cause melanin dispersion; adenosine triphosphate (ATP) and calcium deficient Ringer solution foster melanin concentration. He feels that (1), the lightening effect of ATP is based on its ability to bind the calcium ions in Ringer solution and at cell surfaces rather than on its energy yielding potentialities and (2), pigment concentration requires more metabolic energy than dispersion. This view of the mode of action of ATP differs from that of Lerner and Takahashi (1956) that was discussed in the section on amphibians.

With respect to other groups of reptiles, much less detailed

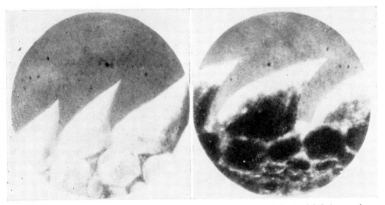

FIG. 40. Marginal dentate scales from pale (left) and dark (right) specimens of *Phrynosoma*. (From Parker, 1938b.)

information is available. Young alligators possess melanophores in the skin of the throat and of the ventral abdomen (Kleinholz, 1941). These melanophores appear to be functional, responding to black and to white backgrounds. Epinephrine concentrates the melanin and MSH disperses it.

Rahn (1941) studied the color changes of snakes, primarily the rattlesnake *Crotalus viridis*. Two types of melanophores were found. One was relatively small and resided primarily in the epidermis; the other was much larger and retained its cell body in the dermis, but sent its processes into the epidermal layer. Rahn found that hypophysectomized specimens remained permanently pale but MSH dispersed the melanin of hypophysectomized rattlesnakes. Preliminary experiments on the garter snake (*Thamnophis ordinoides*), ribbon snake (*Thamnophis radix*), bull snake (*Pituophis sayi*), and Florida water snake (*Natrix sipedon*) indicated that the color change system in these snakes is essentially the same as in the rattlesnake.

Finally, with respect to chelonians, Woolley (1957) reported that *Chelodina longicolis* has melanophores whose pigment concentrated on a white background and dispersed on a black one. However, 30 days were required for each process. MSH dispersed the melanin but epinephrine had no effect. In view of the long time, 30 days, required for either dispersion or concentration of the melanin, the possibility exists that the response Woolley observed was a morphological rather than a physiological color change.

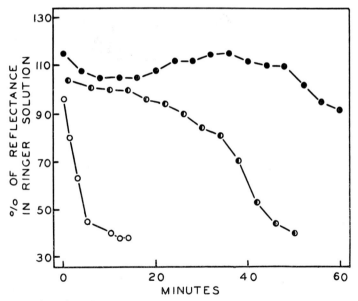

FIG. 41. Dispersion of melanin in skin of *Anolis carolinensis* exposed to MSH. Circles, no pre-treatment; half-filled circles, pretreatment with reduced glutathione for 20 min; dots, pretreatment with cysteine for 20 min. (Redrawn from Horowitz, 1957.)

CHAPTER 7

PERSPECTIVES

DURING the last 15 years much has been accomplished by comparative physiologists who have concentrated on color changes in invertebrates. Consequently, there is no longer the great disparity between the invertebrates and the vertebrates that existed when Parker (1948) published his treatise. However, much still remains to be accomplished with all groups of animals that possess chromatophores. It is apparent to workers in this area that for most of the animals that have been studied we have merely a sketchy outline of their color change systems. Many basic questions still await answers. An attempt was made to indicate in the foregoing chapters those areas where deficiencies lie. For example, we would like to know how the chromatophorotropins stimulate pigment migration, how the pigment migrates, the relationship between neurosecretory granules in arthropods and the circulating form of the various chromatophorotropins, and the chemistry of arthropod chromactivators. Many more species should be investigated in the hope that someday we might be able to understand how color change systems evolved, with particular attention to the evolution of nervous and endocrine control mechanisms. The current high interest in this field gives promise for the future.

REFERENCES

ABOLIN, L. (1925) Beeinflussung des Fischfarbwechsels durch Chemikalien. I. Infundin- und Adrenalinwirkung auf die Melano- und Xanthophoren der Elritze (*Phoxinus laevis* Ag.), *Arch. mikr. Anat.* **104**, 667.

ABRAMOWITZ, A. A. (1936) Physiology of the melanophore system in the catfish, *Ameiurus*, *Biol. Bull.*, *Woods Hole.* **71**, 259.

ABRAMOWITZ, A. A. (1937a) The chromatophorotropic hormone of the Crustacea: standardization, properties and physiology of the eye-stalk glands, *Biol. Bull.*, *Woods Hole.* **72**, 344.

ABRAMOWITZ, A. A. (1937b) The role of the hypophyseal melanophore hormone in the chromatic physiology of *Fundulus*, *Biol. Bull.*, *Woods Hole.* **73**, 134.

ABRAMOWITZ, A. A. (1939) The pituitary control of chromatophores in the dogfish, *Amer. Nat.* **73**, 208.

ABRAMOWITZ, A. A. (1940) Purification of the chromatophorotropic hormone of the crustacean eyestalk, *J. Biol. Chem.* **132**, 501.

ABRAMOWITZ, A. A., and R. K. ABRAMOWITZ (1938) On the specificity and related properties of the crustacean chromatophorotropic hormone, *Biol. Bull.*, *Woods Hole.* **74**, 278.

ALLEN, B. M. (1916) The results of extirpation of the anterior lobe of the hypophysis and of the thyroid of *Rana pipiens* larvae, *Science* **44**, 755.

ARMITAGE, K. B. (1960) Chromatophore behavior of the isopod *Ligia occidentalis* Dana, 1853, *Crustaceana* **1**, 193.

BAGNARA, J. T. (1957) Hypophysectomy and the tail darkening reaction in *Xenopus*, *Proc. Soc. Exper. Biol. N.Y.* **94**, 572.

BAGNARA, J. T. (1960a) Tail melanophores of *Xenopus* in normal development and regeneration, *Biol. Bull.*, *Woods Hole.* **118**, 1.

BAGNARA, J. T. (1960b) Pineal regulation of the body lightening reaction in amphibian larvae, *Science* **132**, 1481.

BAGNARA, J. T. (1961) Onset of pineal and hypophyseal regulation of melanophores in *Xenopus*, *Amer. Zoologist* **1**. 339.

BAGNARA, J. T., and S. NEIDLEMAN (1958) Effect of chromatophorotropic hormone on pigments of anuran skin, *Proc. Soc. Exper. Biol.*, *N.Y.* **97**, 671.

BALLOWITZ, E. (1914) Die chromatischen Organe, Melaniridosomen, in der Haut der Barsche (*Perca* und *Acerina*), *Z. wiss. Zool.* **110**, 1.

BARGMANN, W. (1958) Elektronenmikroskopische Untersuchungen an der Neurohypophyse, *Zweites Internat. Sympos. über Neurosekretion*, W. BARG-

MANN, B. HANSTRÖM, B. SCHARRER, and E. SCHARRER, eds. Springer-Verlag, Berlin.

BARNES, H., and J. J. GONOR (1958) Neurosecretory cells in some cirripedes, *Nature, Lond.* **181**, 194.

BARNOTHY, J., and M. FORRÓ (1939) Diurnal and sidereal effects and the meteorologic influences on shower and vertical intensity of cosmic rays, *Phys. Rev.* **55**, 868.

BLASCHKO, H., P. HAGEN, and A. D. WELCH (1955) Observations on the intracellular granules of the adrenal medulla, *J. Physiol.* **129**, 27.

BLASCHKO, H., and A. D. WELCH (1953) Localization of adrenaline in cytoplasmic particles of the bovine adrenal medulla, *Arch. exp. Path. Pharmak.* **219**, 17.

BLES, E. J. (1905) The life history of *Xenopus laevis* Daud, *Trans. Roy. Soc. Edinb.* **41**, 789.

BLISS, D. E. (1951) Metabolic effect of sinus gland or eyestalk removal in the land crab, *Gecarcinus lateralis, Anat. Rec.* **111**, 502.

BLISS, D. E., J. B. DURAND, and J. H. WELSH (1954) Neurosecretory systems in decapod Crustacea, *Z. Zellforsch.* **39**, 520.

BLISS, D. E., and J. H. WELSH (1952) The neurosecretory system of brachyuran Crustacea, *Biol. Bull., Woods Hole.* **103**, 157.

BORS, O., and W. C. RALSTON (1951) A simple assay of mammalian pineal extracts, *Proc. Soc. exp. Biol., N.Y.* **77**, 807.

BOWMAN, T. E. (1942) Morphological color change in the crayfish, *Amer. Nat.* **76**, 332.

BOWMAN, T. E. (1949) Chromatophorotropins in the central nervous organs of the crab, *Hemigrapsus oregonensis, Biol. Bull., Woods Hole.* **96**, 238.

BOZLER, E. (1928) Über die Tätigkeit der einzelnen glatten Muskelfaser bei der Kontraktion. II. Mitteilung: Die Chromatophorenmuskeln der Cephalopoden, *Z. vergl. Physiol.* **7**, 379.

BOZLER, E. (1931) Über die Tätigkeit der einzelnen glatten Muskelfaser bei der Kontraktion, 3. Mitteilung: Registrierung der Kontraktion der Chromatophorenmuskelzellen von Cephalopoden, *Z. vergl. Physiol.* **13**, 762.

BRANTNER, G. (1956) Die Unabhängigkeit des morphologischen Farbwechsels vom physiologischen Farbwechsel bei der Entstehung des Hochzeitkleides des männlichen Bitterlings, *Z. vergl. Physiol.* **33**, 324.

BREDER, C. M., JR., and P. RASQUIN (1950) A preliminary report on the role of the pineal organ in the control of pigment cells and light reactions in recent teleost fishes, *Science* **111**, 10.

BREDER, C. M., JR., and P. RASQUIN (1955) Further notes on the pigmentation behavior of *Chaetodipterus* in reference to background and water transparency, *Zoologica* **40**, 85.

BRISENO CASTREJON, B., and I. STEVENS FLORES (1955) Cambios producidas por la hipofisectomia en el sistema chromatosico de *Carassius auratus* L., *Anu. Escuela Nacl. Cienc. Biol. (Mex.)* **8**, 203.

BROCH, E. S. (1960) Endocrine control of the chromatophores of the zoeae of the prawn, *Palaemonetes vulgaris, Biol. Bull., Woods Hole* **119**, 305.

BROWN, F. A., JR. (1933) The controlling mechanism of chromatophores in *Palaemonetes, Proc. Nat. Acad. Sci., Wash.* **19**. 327.

BROWN, F. A., JR. (1934) The chemical nature of the pigments and the transformation responsible for color changes in *Palaemonetes, Biol. Bull., Woods Hole.* **67**, 365.

BROWN, F. A., JR. (1935a) Color changes in *Palaemonetes, J. Morph.* **57**, 317.

BROWN, F. A., JR. (1935b) Control of pigment migration within the chromatophores of *Palaemonetes vulgaris, J. Exp. Zool.* **71**, 1.

BROWN, F. A., JR. (1936) Light intensity and melanophore responses in the minnow, *Ericymba buccata* Cope, *Biol. Bull., Woods Hole.* **70**, 8.

BROWN, F. A., JR. (1939a) Background selection in crayfishes, *Ecology* **20**, 507.

BROWN, F. A., JR. (1939b) Humoral control of crustacean chromatophores, *Amer. Nat.* **73**, 247.

BROWN, F. A., JR. (1940) The crustacean sinus gland and chromatophore activation, *Physiol. Zool.* **13**, 343.

BROWN, F. A., JR. (1946) The source and activity of *Crago*-darkening hormone (CDH), *Physiol. Zool.* **19**, 215.

BROWN, F. A., JR. (1948a) Hormones in crustaceans. *The Hormones.* Vol. I, G. PINKUS and K. V. THIMANN eds., Academic Press Inc., New York.

BROWN, F. A., JR. (1948b) Color changes in the stomatopod crustacean. *Chloridella empusa, Anat. Rec.* **101**, 732.

BROWN, F. A., JR. (1950) Studies on the physiology of *Uca* red chromatophores, *Biol. Bull, Woods Hole.* **98**, 218.

BROWN, F. A., JR. (1952a) Hormones in crustaceans. *The Actions of Hormones in Plants and Invertebrates*, K. V. THIMANN, ed., Academic Press, Inc. New York.

BROWN, F. A., JR. (1952b) Chromatophores and color change, *Comparative Animal Physiology*, C. L. PROSSER, ed., W. B. SAUNDERS Co., Philadelphia

BROWN, F. A., JR. (1957) The rhythmic nature of life, *Recent Advances in Invertebrate Physiology*, Univ. Oregon Press, Eugene.

BROWN, F. A., JR. (1958) Studies of the timing mechanism of daily, tidal, and lunar periodicities in organisms, *Perspectives in Marine Biology*, Univ. Calif. Press, Berkeley.

BROWN, F. A., JR., M. F. BENNETT, and C. L. RALPH (1955) Apparent reversible influence of cosmic-ray-induced showers upon a biological system, *Proc. Soc. Exp. Biol., N. Y.* **89**, 332.

BROWN, F. A., JR., and O. CUNNINGHAM (1941) Upon the presence and distribution of a chromatophorotropic principle in the central nervous system of *Limulus, Biol. Bull., Woods Hole.* **81**, 80.

BROWN, F. A., JR., and H. E. EDERSTROM (1940) Dual control of certain black chromatophores of *Crago, J. Exp Zool.* **85**, 53.

BROWN, F. A., JR., and M. FINGERMAN (1951) Differentation of black- and red-dispersing factors from the brain of the fiddler crab, *Uca, Fed. Proc.* **10**, 20.

BROWN, F. A., JR., M. FINGERMAN, and M. N. HINES (1954) A study of the mechanism involved in shifting of the phases of the endogenous daily rhythm by light stimuli, *Biol. Bull., Woods Hole.* **106**, 308.

BROWN, F. A., JR., M. FINGERMAN, M. I. SANDEEN, and H. M. WEBB (1953) Persistent diurnal and tidal rhythms of color change in the fiddler crab, *Uca pugnax, J. Exp. Zool.* **123**, 29.

BROWN, F. A., JR., and M. N. HINES (1952) Modifications in the diurnal pigmentary rhythm of *Uca* effected by continuous illumination, *Physiol. Zool.* **25**, 56.

BROWN, F. A., JR., and I. M. KLOTZ (1947) Separation of two mutually antagonistic chromatophorotropins from the tritocerebral commissure of *Crago, Proc. Soc. Exp. Biol., N. Y.* **64**, 310.

BROWN, F. A., JR., and A. MEGLITSCH (1940) Comparison of the chromatophorotropic activity of insect corpora cardiaca with that of crustacean sinus glands, *Biol. Bull., Woods Hole.* **79**, 409.

BROWN, F. A., JR., and L. M. SAIGH (1946) The comparative distribution of two chromatophorotropic hormones (CDH and CBLH) in crustacean nervous systems, *Biol. Bull., Woods Hole.* **91**, 170.

BROWN, F. A., JR., and M. I. SANDEEN (1948) Responses of the chromatophores of the fiddler crab, *Uca*, to light and temperature, *Physiol. Zool.* **21**, 361.

BROWN, F. A., JR., M. I. SANDEEN, and H. M. WEBB (1948) The influence of illumination on the chromatophore system of *Palaemonetes vulgaris, Anat. Rec.* **101**, 733.

BROWN, F. A., JR., and H. H. SCUDAMORE (1940) Differentiation of two principles from the crustacean sinus gland, *J. Cell. Comp. Physiol.* **15**, 103.

BROWN, F. A., JR., and G. C. STEPHENS (1951) Studies of the daily rhythmicity of the fiddler crab, *Uca*, Modifications by photoperiod, *Biol. Bull., Woods Hole.* **101**, 71.

BROWN, F. A., JR., and D. H. THOMPSON (1937) Melanin dispersion and choice of background in fishes, with special reference to *Ericymba buccata, Copeia.* **1937**, 172.

BROWN, F. A., JR., and H. M. WEBB (1948) Temperature relations of an endogenous daily rhythmicity in the fiddler crab, *Uca, Physiol. Zool.* **21**, 371.

BROWN, F. A., JR., and H. M. WEBB (1949) Studies of the daily rhythmicity of the fiddler crab, *Uca*, Modifications by light, *Physiol. Zool.* **22**, 136.

BROWN, F. A., JR., H. M. WEBB, and M. F. BENNETT (1955) Proof for an endogenous component in persistent solar and lunar rhythmicity in organisms, *Proc. Nat. Acad. Sci., Wash.* **41**, 93.

BROWN, F. A., JR., H. M. WEBB, M. F. BENNETT, and M. I. SANDEEN (1954). Temperature independence of the frequency of the endogenous tidal rhythm of *Uca, Physiol. Zool.* **27**, 345.

BROWN, F. A., JR., H. M. WEBB, and M. I. SANDEEN (1952) The action of two hormones regulating the red chromatophores of *Palaemonetes, J. Exp. Zool.* **120**, 391.

BROWN, F. A., JR., and V. J. WULFF (1941) Chromatophore types in *Crago* and their endocrine control, *J. Cell. Comp. Physiol.* **18**, 339.

BRÜCKE, E. (1852) Untersuchungen über den Farbwechsel des afrikanischen Chamaleons, *Denkschr. Akad. Wiss. Wien.* **4**, 179.

BURGERS, A. C. J. (1956) Investigations into the action of certain hormones and other substances on the melanophores of the South African clawed toad, *Xenopus laevis*, G. W. van der Wiel and Co., Arnheim, Netherlands.

BURGERS, A. C. J. (1958) The reactions of the melanophores of the fiddler crab, *Uca rapax*, to crustacean eyestalk extracts and to vertebrate melanophorotropic hormones, *Proc. Koninkl. Ned. Akad. Wet.* Ser. C, **61**, 597.

BURGERS, A. C. J. (1959) Investigations of the action of hormones and light. on the erythromelanosomes of the swimming crab, *Macropipus vernalis, Pubbl. Staz. Zool. Napoli* **31**, 139.

BURGERS, A. C. J. (1961) Occurrence of three electrophoretic components with melanocyte-stimulating activity in extracts of single pituitary glands from ungulates, *Endocrinology.* **68**, 698.

BURGERS, A. C. J., TH. A. C. BOSCHMAN, and J. C. VAN DE KAMER (1953) Excitement darkening and the effect of adrenaline on the melanophores of *Xenopus laevis, Acta Endocrinol.* **14**, 72.

BURGERS, A. C. J., W. LEEMREIS, T. DOMINICZAK, and C. J. VAN OORDT (1958) Inhibition of the secretion of intermedine by d-lysergic acid diethylamide (LSD 25) in the toad *Xenopus laevis, Acta Endocrinol.* **29**, 191.

BURGERS, A. C. J., and G. J. VAN OORDT (1956) The effect of the skin secretion of *Xenopus laevis* on its dermal melanophores, *Acta Endocrinol,* **23**, 265.

BUTCHER, E. O., and H. B. ADELMANN (1937) The effects of covering and rotating the eyes on the melanophoric responses in *Fundulus heteroclitus, Bull. Mt. Desert Isl. Biol. Lab.,* pp. 16–18.

CARLISLE, D. B. (1953a) Studies on *Lysmata seticaudata* Risso (Crustacea Decapoda). VI. Notes on the structure of the neurosecretory system of the eyestalk, *Publ. Staz. Zool. Napoli* **24**, 435.

CARLISLE, D. B. (1953b) Note préliminaire sur la structure du système neurosécréteur du pédoncule oculaire de *Lysmata seticaudata* Risso (Crustacea), *C. R. Acad. Sci., Paris* **236**, 2541.

CARLISLE, D. B. (1953c) Studies on *Lysmata seticaudata* Risso (Crustacea Decapoda). IV. On the site of origin of the moult-accelerating principle-experimental evidence, *Publ. Staz. Zool., Napoli* **24**, 285.

CARLISLE, D. B. (1954) On the hormonal inhibition of moulting in decapod Crustacea, *J. Mar. Biol. Ass. U. K.* **33**, 61.

CARLISLE, D. B. (1955) Local variations in the colour pattern of the prawn *Leander serratus* (Pennant), *J. Mar. Biol. Ass. U. K.* **34**, 559.

CARLISLE, D. B., M. DUPONT-RAABE, and F. G. W. KNOWLES (1955) Recherches préliminaires relatives à la séparation et à la comparaison des substances chromactives des Crustacés et des Insectes, *C. R. Acad. Sci., Paris* **240**, 665.

CARLISLE, D. B., and F. G. W. KNOWLES (1953) Neurohaemal organs in crustaceans, *Nature, Lond.* **172**, 404.

CARLISLE, D. B., and F. G. W. KNOWLES (1959) *Endocrine Control in Crustaceans*, Cambridge University Press, Cambridge.

CARLISLE, D. B., and L. M. PASSANO (1953) The X-organ of Crustacea, *Nature, Lond.* **171**, 1070.

CARLSON, S. PH. (1935) The color changes in *Uca pugilator, Proc. Nat. Acad. Sci., Wash.* **21**, 549.

CARLSON, S. PH. (1936) Colour changes in Brachyura crustaceans, especially in *Uca pugilator, Kungl. Fysiogr. Sälls. Lund. Forhandl.* **6**, 63.

CARLSSON, A. N., A. HILLARP, and B. HÖKFELT (1957) The concomitant release of adenosine triphosphate and catechol amines from the adrenal medulla, *J. Biol. Chem.* **227**, 243.

CARSTAM. S. PH. (1951) Enzymatic inactivation of the pigment hormone of the crustacean sinus gland, *Nature, Lond.* **167**, 321.

CARSTAM, S. PH., and S. SUNESON (1949) Pigment activation in *Idothea neglecta* and *Leander adspersus, Kungl. Fysiogr. Sälls. Lund. Forhandl.* **19**, 1.

CHANG, C. Y. (1957) Thyroxine effect on melanophore contraction in *Xenopus laevis, Science* **126**, 121.

COONFIELD, B. R. (1940) The pigment in the skin of *Myxine glutinosa* Linn., *Trans. Amer. Micr. Soc.* **59**, 398.

DAVEY, K. G. (1959) Serotonin and change of colour in frogs, *Nature, Lond.*, **183**, 1271.

DAVEY, K. G. (1960) Intermedin and change of color in frogs: a new hypothesis, *Canad. J. Zool.* **38**, 715.

DAWSON, A. B. (1953) Evidence for the termination of neurosecretory fibers within the pars intermedia of the hypophysis of the frog, *Rana pipiens, Anat. Rec.* **115**, 63.

DEANIN, G. G., and F. R. STEGGERDA (1948) Use of the spectrophotometer for measuring melanin dispersion in the frog, *Proc. Soc. Exp. Biol., N.Y.* **67**, 101.

DEUTSCH, S., E. T. ANGELAKOS, and E. R. LOEW (1957) Comparison of methods for quantitating melanophore responses, *Proc. Soc. Exp. Biol.*, *N.Y.* **94**, 576.

DRACH, P. (1944) Étude préliminaire sur le cycle d'intermue et son conditionnement hormonal chez *Leander serratus*, *Bull, Biol. de France et de Belgique,* **78**, 40.

DUELOFEU, V. (1935) Adrenalin in Gift von *Bufo arenarum*, *Hoppe-Seyl. Z.* **237**, 171.

DUNCAN, D. (1955) Electron microscopy of the hypophysis, pars neuralis, *Anat. Rec.* **121**, 430.

DUNCAN, D. (1956) An electron microscope study of the neurohypophysis of a bird, *Gallus domesticus*, *Anat. Rec.* **125**, 457.

DUPONT-RAABE, M. (1949) Réactions humorales des chromatophores de la larve de Corèthre, *C. R. Acad. Sci.*, *Paris* **228**, 130.

DUPONT-RAABE, M. (1951) Étude expérimentale de l'adaptation chromatique chez le phasme, *Carausius morosus*, *C. R. Acad. Sci.*, *Paris* **232**, 886.

DUPONT-RAABE, M. (1954) Le rôle endocrine du cerveau dans la régulation des phènoménes d'adaptation chromatique et de la ponte chez les Phasmides, *Publ. Staz. Zool. Suppl.* **24**, 62.

DUPONT-RAABE, M. (1956a) Mise en évidence sur les phasmides d'une troisième paire de *nervi corporis cardiaci*, voie possible de cheminement de la substance chromactive tritocérébrale vers les Corpora cardiaca, *C. R. Acad. Sci.*, *Paris* **243**, 1240.

DUPONT-RAABE, M. (1956b) Rôle des différents éléments du système nerveux central dans la variation chromatique des phasmides, *C. R. Acad. Sci.*, *Paris* **243**, 1358.

DUPONT-RAABE, M. (1958) Quelques aspects des phénomènes de neurosécrétion chez les Phasmides, *Zweiter Internat. Symp. Neurosekretion*, W. BARGMANN, B. HANSTRÖM, B. SCHARRER, and E. SCHARRER, eds. SPRINGER-VERLAG, Berlin.

DURAND, J. B. (1956) Neurosecretory cell types and their secretory activity in the crayfish, *Biol. Bull., Woods Hole.* **111**, 62.

EDGREN, R. A. (1954a) Factors controlling color change in the tree frog, *Hyla versicolor* Wied, *Proc. Soc. Exp. Biol., N. Y.* **87**, 20.

EDGREN, R. A. (1954b) Use of *Hyla cinerea* in assay of melanophorotropic potency of ACTH, *Proc. Soc. Exp. Biol., N. Y.* **85**, 229.

EDMAN, P., R. FÄNGE, and E. ÖSTLUND (1958) Isolation of the red pigment concentrating hormone of the crustacean eyestalk, *Zweites Internat. Symp. über Neurosekretion*, W. BARGMANN, B. HANSTRÖM, B. SCHARRER, and E. SCHARRER, eds., SPRINGER-VERLAG, Berlin.

ENAMI, M. (1939) Post-mortem pigmentary reaction in a sea catfish, *Plotosus anguillaris, Proc. Imp. Acad., Tokyo* **15**, 230.

ENAMI, M. (1940) Action mélano-dilatatrice de l'adrénaline chez un silure chat, *Parasilurus asotus, Proc. Imp. Acad., Tokyo* **16**, 236.

ENAMI, M. (1941a) Melanophore response in an isopod crustacean, *Ligia exotica*, I. General responses, *Jap. J. Zool.* **9**, 497.

ENAMI, M. (1941b) Melanophore response in an isopod crustacean, *Ligia exotica*, II. Humoral control of melanophores, *Jap. J. Zool.* **9**, 515.

ENAMI, M. (1943) Chromatophore activator in the central nervous organs of *Uca dubia*, *Proc. Imp. Acad., Tokyo* **18**, 693.

ENAMI, M. (1949) Studies on the controlling mechanism of black chromatophores in the young of a fresh-water crab, *Sesarma haematocheir*, I. On a humoral principle from several ganglionic tissues as concerned with the pigmentary activities, *Physiol. and Ecology* (Kyoto, Japan) **3**, 23.

ENAMI, M. (1951a) The sources and activities of two chromatophorotropic hormones in crabs of the genus *Sesarma*, II. Histology of incretory elements, *Biol. Bull., Woods Hole.* **101**, 241.

ENAMI, M. (1951b) The sources and activities of two chromatophorotropic hormones in crabs of the genus *Sesarma*, I. Experimental Analysis, *Biol. Bull., Woods Hole.* **100**, 28.

ENAMI, M. (1955) Melanophore contracting hormone (MCH) of possible hypothalamic origin in the catfish, *Parasilurus*, *Science* **121**, 36.

ERGENE, S. (1950) Untersuchungen über Farbanpassung und Farbwechsel bei *Acrida turrita*, *Z. vergl. Physiol.* **32**, 530.

ERGENE, S. (1952) Spielt das Auge beim homochromen Farbwechsel von *Acrida turrita* eine Rolle? *Z. vergl. Physiol.* **34**, 159.

ERGENE, S. (1953) Homochrome Farbanpassungen bei *Mantis religiosa*, *Z. vergl. Physiol.* **35**, 36.

ERGENE, S. (1954) Über den angeblichen Einfluss von frischem grünen Futter auf den Farbwechsel von *Acrida turrita*, *Z. vergl. Physiol.* **36**, 235.

ERGENE, S. (1955a) Über die Faktoren, die Grünfärbung bei *Acrida* bedingen, *Z. vergl. Physiol.* **37**, 221.

ERGENE, S. (1955b) Farbe der Hämolymphe vor und nach dem Farbwechsel (Untersuchungen an *Mantis religiosa*), *Istanbul Univ. Fen. Fak. Mecmausi, Ser. B, Sci. Nat.* **20**, 109.

ERGENE, S. (1955c) Der imaginale homochromer Farbwechsel bei Orthopteren, *Istanbul Univ. Fen. Fak. Mecmausi. Ser. B, Sci. Nat.* **20**, 113.

ERGENE, S. (1955d) Weitere Untersuchungen über Farbanpassung bei *Oedaleus decorus*, *Z. vergl. Physiol.* **37**, 226.

ERGENE, S. (1956a) Homochromer Farbwechsel und Verhalten zur Bodenfarbe bei *Tylopsis liliifolia*-Imagines, *Istanbul Univ. Fen. Fak. Mecmausi, Ser. B, Sci. Nat.* **21**, 61.

ERGENE, S. (1956b) Homochromer Farbwechsel bei geblendeten *Oedaleus decorus*, *Z. vergl. Physiol.* **38**, 311.

ERGENE, S. (1956c) Farbanpassung bei *Tylopsis liliifolia*-larven, *Z. vergl. Physiol.* **38**, 315.

ERSPAMER, V. (1954) Quantitative estimation of 5-HT in vertebrates, In *CIBA*

Foundation Symposium on Hypertension: Humoral and Neurogenic Factors J. and A. Churchill Ltd., London.

ERSPAMER, V., and B. ASERO (1953) Isolation of enteramine from extracts of posterior salivary glands of *Octopus vulgaris* and of *Discoglossus pictus* skin, *J. Biol. Chem.* **200**, 311.

FALK, S., and J. RHODIN (1957) Mechanism of pigment migration within teleost melanophores, In *Electron Microscopy: Proceedings of the Stockholm Conference*, F. S. SJÖSTRAND and J. RHODIN eds., Academic Press Inc., New York.

FINGERMAN, M. (1955) Persistent daily and tidal rhythms of color change in *Callinectes sapidus, Biol. Bull., Woods Hole.* **109**, 255.

FINGERMAN, M. (1956a) Physiology of the black and red chromatophores of *Callinectes sapidus, J. Exp. Zool.* **133**, 87.

FINGERMAN, M. (1956b) The physiology of the melanophores of the isopod *Ligia exotica, Tulane Studies Zool.* **3**, 137.

FINGERMAN, M. (1956c) Black pigment concentrating factor in the fiddler crab, *Science* **123**, 585.

FINGERMAN, M. (1956d) Phase difference in the tidal rhythms of color change of two species of fiddler crab, *Biol. Bull., Woods Hole.* **110**, 274.

FINGERMAN, M. (1957a) Endocrine control of the red and white chromatophores of the dwarf crawfish, *Cambarellus shufeldti, Tulane Studies Zool.* **5**, 137.

FINGERMAN, M. (1957b) Lunar rhythmicity in marine organisms, *Amer. Nat.* **111**, 167.

FINGERMAN, M. (1957c) Relation between position of burrows and tidal rhythm of *Uca, Biol. Bull., Woods Hole.* **112**, 7.

FINGERMAN, M. (1957d) Physiology of the red and white chromatophores of the dwarf crayfish, *Cambarellus shufeldti, Physiol. Zool.* **30**, 142.

FINGERMAN, M. (1958) The chromatophore system of the crawfish *Orconectes clypeatus, Amer. Midl. Nat.* **60**, 71.

FINGERMAN, M. (1959a) The physiology of chromatophores, *Int. Rev. Cyt.* **8**, 175.

FINGERMAN, M. (1959b) Physico-chemical characterization of chromatophorotropins in the crayfish *Cambarellus shufeldti, Biol. Bull., Woods Hole.* **117**, 382.

FINGERMAN, M. (1959c) Comparison of the chromatophorotropins of two crayfishes with special reference to electrophoretic behavior, *Tulane Studies Zool.* **7**, 21.

FINGERMAN, M. (1960) Tidal rhythmicity in marine organisms, *Cold Spr. Harb. Symp. Quant. Biol.* **25**, 481.

FINGERMAN, M., and T. AOTO (1958a) Electrophoretic analysis of chromatophorotropins in the dwarf crayfish, *Cambarellus shufeldti, J. Exp. Zool.* **138**, 25.

FINGERMAN, M., and T. AOTO (1958b) Chromatophorotropins in the crayfish

Orconectes clypeatus and their relationship to long-term background adaptation, *Physiol. Zool.* **31**, 193.

FINGERMAN, M., and T. AOTO (1959) The neurosecretory system of the dwarf crayfish, *Cambarellus shufeldti*, revealed by electron and light microscopy, *Trans. Amer. Micr. Soc.* **78**, 305.

FINGERMAN, M., and T. AOTO (1960) Effects of eyestalk ablation upon neurosecretion in the supraesophageal ganglia of the dwarf crayfish *Cambarellus shufeldti*, *Trans. Amer. Micr. Soc.* **79**, 68.

FINGERMAN, M., and C. FITZPATRICK (1956) An endocrine basis for the sexual difference in melanin dispersion in *Uca pugilator*, *Biol. Bull., Woods Hole*. **110**, 138.

FINGERMAN, M., and M. E. LOWE (1957a) Influence of time on background upon the chromatophore system of two crustaceans, *Physiol. Zool.* **30**, 216.

FINGERMAN, M., and M. E. LOWE (1957b) Hormones controlling the chromatophores of the dwarf crayfish, *Cambarellus shufeldti*: their secretion, stability, and separation by filter paper electrophoresis, *Tulane Studies Zool.* **5**, 151.

FINGERMAN, M., and M. E. LOWE (1958) Stability of the chromatophorotropins of the dwarf crayfish, *Cambarellus shufeldti*, and their effects on another crayfish, *Biol. Bull., Woods Hole*. **114**, 317.

FINGERMAN, M., M. E. LOWE, and W. C. MOBBERLY, JR. (1958) Environmental factors involved in setting the phases of tidal rhythm of color change in the fiddler crabs *Uca pugilator* and *Uca minax*, *Limnol. and Oceanog.* **3**, 271.

FINGERMAN, M., and W. C. MOBBERLY, JR. (1960a) Physicochemical properties and differentiation of chromatophorotropins and retinal pigment light-adapting hormone in the dwarf crayfish, *Cambarellus shufeldti*, *Amer. Midl. Nat.* **64**, 474.

FINGERMAN, M., and W. C. MOBBERLY, JR. (1960b) Differentiation of chromatophorotropins and retinal pigment light-adapting hormone from the eyestalk of the dwarf crayfish, *Cambarellus shufeldti*, *Bull. Ass. Southeastern Biologists* **7**, 27.

FINGERMAN, M., and W. C. MOBBERLY, JR. (1960c) Trophic substances in a blind cave crayfish, *Science* **132**, 44.

FINGERMAN, M., R. NAGABHUSHANAM, and L. PHILPOTT (1960a) Responses of the melanophores of the grapsoid crab *Sesarma reticulatum* to light and temperature, *Biol. Bull., Woods Hole*. **119**, 315.

FINGERMAN, M., R. NAGABHUSHANAM, and L. PHILPOTT (1960b) The influence of light and endocrines on the chromatophores of the mud shrimp, *Upogebia affinis*, *Biol. Bull., Woods Hole*. **119**, 314.

FINGERMAN, M., R. NAGABHUSHANAM, and L. PHILPOTT (1960c) The responses of the melanophores of eyestalkless specimens of *Sesarma reticulatum* to illumination and endocrines, *Biol. Bull., Woods Hole*. **119**, 315.

FINGERMAN, M., R. NAGABHUSHANAM, and L. PHILPOTT (1961) Physiology of the melanophores in the crab *Sesarma reticulatum*, *Biol. Bull.*, *Woods Hole.* **120**, 326.

FINGERMAN, M., M. I. SANDEEN, and M. E. LOWE (1959) Experimental analysis of the red chromatophore system of the prawn *Palaemonetes vulgaris*, *Physiol. Zool.* **32**, 128.

FINGERMAN, M., and D. W. TINKLE (1956) Responses of the white chromatophores of two species of prawns (*Palaemonetes*) to light and temperature, *Biol. Bull.*, *Woods Hole.* **110**, 144.

FORGÁCS, P. (1956) Corticotrophin and melanophore expanding hormone, *Acta Endocrinol.* **23**, 1.

FORRÓ, M. (1937) Diurnal variation of cosmic ray shower, *Nature, Lond.* **139**, 633.

FOX, D. L. (1953) *Animal Biochromes and Structural Colours*, Cambridge Univ. Press, Cambridge.

FOX, H. M., and G. VEVERS (1960) *The Nature of Animal Colours*, SIDGWICK and JACKSON, LTD., London.

FRIEDEN, E. H., and J. M. BOZER (1951) Effect of administration of intermedin upon melanin content of the skin for *Rana pipiens*, *Proc. Soc. Exp. Biol.*, *N.Y.* **77**, 35.

FROST, R., R. SALOUN, and L. H. KLEINHOLZ (1951) Effect of sinus gland and of eyestalk removal on rate of oxygen consumption in *Astacus*, *Anat. Rec.* **111**, 572.

FUCHS, R. F. (1914) Der Farbenwechsel und die chromatische Hautfunktion der Tiere, H. Winterstein, *Handb. vergl. Physiol.* **3**, 1189.

FUJII, R. (1960) The seat of atropine action in the melanophore-dispersing system of fish, *J. Fac. Sci.*, *Tokyo Univ.* **8**, 643.

FUJITA, H. (1957) Electron microscopic observation on the neurosecretory granules in the pituitary posterior lobe of dog, *Arch. Hist. Jap.* **12**, 165.

GERSCH, M. (1956) Untersuchungen zur Frage der hormonalen Beeinflussung der Melanophoren bei der *Corethra*-larve, *Z. vergl. Physiol.* **39**, 190.

GERSCH, M. (1958) Neurohormone bei wirbellosen Tieren, *Verhandl. Deutsch. Zool. Gesell. in Frankfurt a. M.*, **1958**, 40.

GERSCH, M., F. FISCHER, H. UNGER, and H. KOCH (1960) Die Isolierung neurohormonaler Faktoren aus dem Nervensystem der Kuchenschabe *Periplaneta americana*, *Z. Naturf.* **15**, 319.

GESCHWIND, I. I. (1959) Species variation in protein and polypeptide hormones, *Comparative Endocrinology*, A. GORBMAN, ed., John Wiley and Sons, New York.

GESCHWIND, I. I., and C. H. LI (1957) The isolation and characterization of a melanocyte-stimulating hormone (β-MSH) from hog pituitary glands, *J. Amer. Chem. Soc.* **79**, 615.

GESCHWIND, I. I., C. H. LI, and L. BARNAFI (1956) Isolation and structure of the melanocyte-stimulating hormone from porcine pituitary glands, *J. Amer. Chem. Soc.* **78**, 4494.

GESCHWIND, I. I., C. H. LI, and L. BARNAFI (1957a) The structure of the β-melanocyte-stimulating hormone, *J. Amer. Chem. Soc.* **79**, 620.

GESCHWIND, I. I., C. H. LI, and L. BARNAFI (1957b) The isolation, characterization and amino acid sequence of a melanocyte-stimulating hormone from bovine pituitary glands, *J. Amer. Chem. Soc.* **79**, 6394.

GESCHWIND, I. I., W. O. REINHARDT, and C. H. LI (1952) Observations on the connexion between intermedin and adrenocorticotropic hormone, *Nature, Lond.* **169**, 1061.

GIERSBERG, H. (1928) Über den morphologischen und physiologischen Farbwechsel der Stabheuschrecke *Dixippus* (*Carausius*) *morosus, Z. vergl. Physiol.* **7**, 657.

GIERSBERG, H. (1931) Der Farbwechsel der Fische, *Z. vergl. Physiol.* **13**, 258.

GIRARDIE, A. (1962) Fonctions de la pars intercerebralis chez *Locusta migratoria* L., *C. R. Acad. Sci. Paris* **254** 2669.

GOODWIN, T. W. (1951) Carotenoids in fish, *Biochem. Soc. Symp.* (*Cambridge, England*) **6**, 63.

GOODWIN, T. W. (1960) Biochemistry of pigments, *The Physiology of Crustacea, I. Metabolism and Growth*, T. H. WATERMAN ed., Academic Press Inc., New York.

GRAY, E. G. (1956) Control of the melanophores of the minnow (*Phoxinus phoxinus* (L.)), *J. Exp. Biol.* **33**, 448.

GREEN, J. D. (1951) The comparative anatomy of the hypophysis with special reference to its blood supply and innervation, *Amer. J. Anat.* **88**, 225.

GREEN, J. D., and D. L. Maxwell (1959) Comparative anatomy of the hypophysis and observations on the mechanism of neurosecretion, *Comparative Endocrinology*, A. GORBMAN, ed., John Wiley and Sons, New York.

GRIFFITHS, M. (1936) The colour changes of batoid fishes, *Proc. Linn. Soc. N.S.W.* **61**, 318.

HADLEY, C. E. (1929) Color changes in two Cuban lizards, *Bull. Mus. Comp. Zool. Harvard* **69**, 105.

HADORN, E., and G. FRIZZI (1949) Experimentelle Untersuchungen zur Melanophoren Reaktion von *Corethra, Rev. Suisse Zool.* **56**, 306.

HANSTRÖM, B. (1933) Neue Untersuchungen über Sinnesorgane und Nervensystem der Crustaceen, II. *Zool. Jahrb. Abt. Anat. Ontog. Tiere* **56**, 387.

HANSTRÖM, B. (1937) Die Sinusdrüse und der Hormonale bedingte Farbwechsel der Crustaceen, *Kungl. Svenska Vetensk. Handl.* **16**, 1.

HANSTRÖM, B. (1939) *Hormones in Invertebrates*, Oxford Univ. Press, Oxford.

HANSTRÖM, B. (1948a) Three principal incretory organs in the animal kingdom. The sinus gland in crustaceans, the corpus cardiacum-allatum in insects, the hypophysis in vertebrates, *Bull. Biol. de France et de Belgique, Suppl.* **33**, 182.

HANSTRÖM, B. (1948b) The brain, the sense organs, and the incretory organs of the head in the Crustacea Malacostraca, *Bull. Biol. de France et de Belgique, Suppl.* **33**, 98.

HARKER, J. (1960) Endocrine and nervous factors in insect circadian rhythms, *Cold Spr. Harb. Symp. Quant. Biol.* **25**, 279.

HARRIS, J. I. (1959) Structure of a melanocyte-stimulating hormone from the human pituitary gland, *Nature, Lond.* **184**, 167.

HARRIS, J. I., and A. B. LERNER (1957) Amino-acid sequence of the α-melanocyte-stimulating hormone, *Nature, Lond.* **179**, 1346.

HARRIS, J. I., and P. ROOS (1956) Amino-acid sequence of a melanocyte-stimulating peptide, *Nature, Lond.* **178**, 90.

HAVEL, V. J., and L. H. KLEINHOLZ (1951) Effects of seasonal variation, sinus gland removal, and eyestalk removal on concentration of blood calcium in *Astacus, Anat. Rec.* **111**, 571.

HEALEY, E. G. (1940) Über den Farbwechsel der Elritze (*Phoxinus laevis* Ag.), *Z. vergl. Physiol.* **27**, 545.

HEALEY, E. G. (1951) The colour change of the minnow (*Phoxinus laevis* Ag.), I. Effects of spinal section between vertebrae 5 and 12 on the responses of the melanophores, *J. Exp. Bicl.* **28**, 297.

HEALEY, E. G. (1954) The colour change of the minnow (*Phoxinus laevis* Ag.), II. Effects of spinal section between vertebrae 1 and 15 and of anterior autonomic section on the responses of the melanophores, *J. Exp. Biol.* **31**, 473.

HERRICK, E. H. (1933) The structure of epidermal melanophores in frog tadpoles, *Biol. Bull., Wood's Hole.* **64**, 304.

HILL, A. V., J. L. PARKINSON, and D. Y. SOLANDT (1935) Photoelectric records of the colour change in *Fundulus heteroclitus, J. Exp. Biol.* **12**, 397.

HILLARP, N. A., S. LAGERSTED, and B. NILSON (1953) The isolation of a granular fraction from the suprarenal medulla, containing the sympathomimetic catechol amines, *Acta Physiol. Scand.* **29**, 251.

HILLARP, N. A., and B. NILSON (1954) The structure of the adrenaline and noradrenaline containing granules in the adrenal medullary cells with reference to the storage and release of the sympathomimetic amines, *Acta Physiol. Scand.* **31**, 79.

HINES, M. N. (1954) A tidal rhythm in behavior of melanophores in autotomized legs of *Uca pugnax, Biol. Bull., Woods Hole.* **107**, 386.

HODGE, M. H., and G. B. CHAPMAN (1958) Some observations on the fine structure of the sinus gland of a land crab, *Gecarcinus lateralis, J. Biophys. Biochem. Cyt.* **4**, 571.

HOGBEN, L. T. (1936) The pigmentary effector system. VII. The chromatic function in elasmobranch fishes, *Proc. Roy. Soc. B.* **120**, 142.

HOGBEN, L. T., and F. W. LANDGREBE (1940) The pigmentary effector system. IX. The receptor fields of the teleostean visual response, *Proc. Roy. Soc. B.* **128**, 317.

HOGBEN, L. T., and L. MIRVISH (1928) The pigmentary effector system. V. The nervous control of excitement pallor in reptiles, *J. Exp. Biol.* **5**, 295.

HOGBEN, L. T., and D. SLOME (1931) The pigmentary effector system. VI. The dual character of endocrine coordination in amphibian colour change, *Proc. Roy. Soc., B.* **108**, 10.

HOGBEN, L. T., and D. SLOME (1936) The pigmentary effector system. VIII. The dual receptive mechanism of the amphibian background response, *Proc. Roy. Soc., B.* **120**, 158.

HOGBEN, L. T., and F. R. WINTON (1923) The pigmentary effector system. III. Colour response in the hypophysectomized frog, *Proc. Roy. Soc., B,* **95**, 15.

HOROWITZ, S. B. (1957) The effects of sulfhydryl inhibitors and thiol compounds on pigment aggregation and dispersion in the melanophores of *Anolis carolinensis*, *Exp. Cell Res.* **13**, 400.

HOROWITZ, S. B. (1958) The energy requirements of melanin granule aggregation and dispersion in the melanophores of *Anolis carolinensis*, *J. Cell. Comp. Physiol.* **51**, 341.

HOWARD, K. S., R. G. SHEPHERD, E. A. EIGNER, D. S. DAVIES, and P. H. BELL (1955) Structure of B-corticotropin: final sequence studies, *J. Amer. Chem. Soc.* **77**, 3419.

IMAI, K. (1958) Extraction of melanophore concentrating hormone (MCH) from the pituitary of fishes, *Endocrinologia Japon.* **5**, 34.

JANDA, V. (1934) Contribution à l'étude des changements périodiques de la coloration chez *Dixippus morosus* Br. et Redt., *Mém. Soc. Sci. Bohème,* **1934**, 1.

JOHNSSON, S., and B. HÖGBERG (1952) Observations on the connexion between intermedin and adrenocorticotropic hormone, *Nature, Lond.* **169**, 286.

JOLY, P. (1951) Déterminisme endocrine de la pigmentation chez *Locusta migratoria* L., *C. R. Soc. Biol., Paris* **145**, 1362.

JOLY, P. (1952) Déterminisme de la pigmentation chez *Acrida turrita* L. (Insecte orthoptéroide), *C. R. Acad. Sci., Paris* **235**, 1054.

JØRGENSEN, C. B., and L. O. LARSEN (1960) Control of colour change in amphibians, *Nature, Lond.* **186**, 641.

JØRGENSEN, C. B., L. O. LARSEN, P. ROSENKILDE, and K. G. WINGSTRAND (1960) Effect of extirpation of median emince on function of pars distalis of the hypophysis in the toad *Bufo bufo* (L.), *Comp. Biochem. Physiol.* **1**, 38.

KAHR, H. (1959) Zur endokrinen Steuerung der Melanophoren-Reaktion bei *Octopus vulgaris*, *Z. vergl. Physiol.* **41**, 435.

KAHR, H., and W. FISCHER (1957) Die Wirkung des 5-Oxytryptamine auf das Pigmentsystem der Haut, *Klin. Wchschr.* **35**, 41.

KALMUS, H. (1938) Über einen latenten physiologischen Farbwechsel beim Flusskrebs *Potamobius astacus*, sowie seine hormonale Beeinflussung, *Z. vergl. Physiol.* **25**, 784.

KARKUM, J. N., A. B. KAR, and B. MUKERJI (1953) Evidence against corticotrophin-like action of melanophore hormone on the adrenal cortex of mice, *Acta Endocrinol.* **13**, 188.

KETTERER, B., and E. REMILTON (1954a) Studies on the pituitary melanophore-expanding hormone with reference to its identity with ACTH. II. The effect of heating in alkali on three melanophore-expanding extracts, *J. Endocrin.* **11**, 14.

KETTERER, B., and E. REMILTON (1954b) Studies on the pituitary melanophore-expanding hormone with reference to its identity with ACTH. I. The critical assay of the melanophore-expanding hormone, *J. Endocrin.* **11**, 7.

KEY, K. H. L., and M. F. DAY (1954a) A temperature controlled physiological colour response in the grasshopper *Kosciuscola tristis*, Sjöst. (Orthoptera, Acrididae), *Aust. J. Zool.* **2**, 309.

KEY, K. H. L., and M. F. DAY (1954b) The physiological mechanism of colour change in the grasshopper *Kosciuscola tristis*, Sjöst. (Orthoptera, Acrididae), *Aust. J. Zool.* **2**, 340.

KINOSITA, H. (1953) Studies on the mechanism of pigment migration within fish melanophores with special reference to their electric potentials, *Annot. Zool. Japon.* **26**, 115.

KLEINHOLZ, L. H. (1935) The melanophore-dispersing principle in the hypophysis of *Fundulus heteroclitus*, *Biol. Bull.*, *Woods Hole.* **69**, 379.

KLEINHOLZ, L. H. (1937) Studies in the pigmentary system of Crustacea, I. Color changes and diurnal rhythm in *Ligia baudiniana*, *Biol. Bull.*, *Woods Hole.* **72**, 24.

KLEINHOLZ, L. H. (1938a) Studies in reptilian colour changes, II. The pituitary and adrenal glands in the regulation of the melanophores of *Anolis carolinensis*, *J. Exp. Biol.* **15**, 474.

KLEINHOLZ, L. H. (1938b) Color changes in echinoderms, *Pubbl. Staz. Zool. Napoli* **17**, 53.

KLEINHOLZ, L. H. (1938c) Studies in reptilian colour changes, III. Control of the light phase and behaviour of isolated skin, *J. Exp. Biol.* **15**, 492,

KLEINHOLZ, L. H. (1941) Behavior of melanophores in the alligator, *Anat, Rec.* **81**, Suppl. 121.

KLEINHOLZ, L. H. (1957) Endocrinology of invertebrates, particularly crustaceans. In *Recent Advances in Invertebrate Physiology*, Univ. Oregon Press, Eugene, Oregon.

KLEINHOLZ, L. H., and H. RAHN (1940) The distribution of intermedin:

a new biological method of assay and results of tests under normal and experimental conditions, *Anat. Rec.* **76**, 157.

KNOWLES, F. G. W. (1950) The control of retinal pigment migration in *Leander serratus*, *Biol. Bull.*, *Woods Hole.* **98**, 66.

KNOWLES, F. G. W. (1952) Pigment movements after sinus-gland removal in *Leander adspersus*, *Physiol. Comp. et Oecol.* **2**, 289.

KNOWLES, F. G. W. (1953) Endocrine control in the crustacean nervous system, *Proc. Roy. Soc. B*, **141**, 248.

KNOWLES, F. G. W. (1954) Neurosecretion in the tritocerebral complex of crustaceans, *Pubbl. Staz. Zool. Napoli*, **24**, 74.

KNOWLES, F. G. W. (1956) Some problems in the study of colour change in crustaceans, *Ann. Sci. Nat. Zool. et Biol. Animale* **18**, 315.

KNOWLES, F. G. W. (1958) Electron microscopy of a crustacean neurosecretory organ, *Zweites Internat. Symp. über Neurosekretion*, W. BARGMANN, B. HANSTRÖM, B. SCHARRER, and E. SCHARRER eds., SPRINGER-VERLAG, Berlin.

KNOWLES, F. G. W. (1959) The control of pigmentary effectors, *Comparative Endocrinology*, A. GORBMAN ed., JOHN WILEY and SONS, INC., New York.

KNOWLES, F. G. W., and D. B. CARLISLE (1956) Endocrine control in the Crustacea, *Biol. Rev.*, **31**, 396.

KNOWLES, F. G. W., D. B. CARLISLE, and M. DUPONT-RAABE (1955) Studies on pigment activating substances in animals, I. The separation by paper electrophoresis of chromactivating substances in anthropods, *J. Mar. Biol. Ass.*, U. K., **34**, 611.

KNOWLES, F. G. W., D. B. CARLISLE, and M. DUPONT-RAABE (1956) Inactivation enzymatique d'une substance chromactive des insectes et des crustacés, *C. R. Acad. Sci. Paris*, **242**, 825.

KOBAYASHI, H., S. ISHII, and A. GORBMAN (1959) The hypothalamic neurosecretory apparatus and the pituitary gland of a teleost, *Lepidogobius lepidus*, *Gunma J. Med. Sci.* **8**, 301.

KOHLER, V. (1952) Die Wirkung des Adrenocorticotropins auf die Lipophoren der Pfrille (*Phoxinus laevis*), *Naturwissenshaften* **39**, 554.

KOLLER, G. (1925) Farbwechsel bei *Crangon vulgaris*, *Verh. dtsch. zool. Ges.* **30**, 128.

KOLLER, G. (1927) Über Chromatophorensystem, Farbensinn und Farbwechsel bei *Crangon vulgaris*, *Z. vergl. Physiol.* **5**, 191.

KOLLER, G. (1928) Versuche über die inkretorischen Vorgänge beim Garneelenfarbwechsel, *Z. vergl. Physiol.* **8**, 601.

KOPENEC, A. (1949) Farbwechsel den Larve von *Corethra plumicornis*, *Z. vergl. Physiol.* **31**, 490.

KRAMER, S. (1960) Color changes correlated with parental behavior in cichlid fish, *Anat. Rec.* **138**, 362.

KROPP, B. (1927) The control of the melanophores in the frog, *J. Exp. Zool.* **49**, 289.

KRÖYER, H. (1842) Monographisk Fremstilling auf Slaegten Hippolyte's Nordiske Arter, *K. danske videmsk. Selsk. Skr.* **9**, 209.

KÜHN, A. (1950) Über Farbwechsel und Farbsinn von Cephalopoden, *Z. vergl. Physiol.* **32**, 572.

KULEMANN, H. (1960) Untersuchungen der Pigmentbewegungen in embryonalen Melanophoren von *Xenopus laevis* in Gewebekulturen, *Zool. Jber.* **69**, 169.

LANZING, W. J. R. (1954) The occurrence of a waterbalance, a melanophore-expanding and an oxytocic principle in the pituitary gland of the river-lamprey (*Lampetra fluviatilis* L.), *Acta Endocrinol.* **12**, 277.

LEE, T. H., and A. B. LERNER (1956) Isolation of melanocyte-stimulatin hormone from hog pituitary gland, *J. Biol. Chem.* **221**, 943.

LERNER, A. B. (1961) Hormones and skin color, *Sci. Amer.* **205**, 98.

LERNER, A. B., J. D. CASE, and R. V. HEINZELMAN (1959) Structure of melatonin, *J. Amer. Chem. Soc.* **81**, 6084.

LERNER, A. B., J. D. CASE, and Y. TAKAHASHI (1960) Isolation of melatonin and 5-methoxyindole-3-acetic acid from bovine pineal glands, *J. Biol, Chem.* **235**, 1992.

LERNER, A. B., J. D. CASE, Y. TAKAHASHI, T. H. LEE, and W. MORI (1958) Isolation of melatonin, the pineal gland factor that lightens melanocytes, *J. Amer. Chem. Soc.* **80**, 2587.

LERNER, A. B., and Y. TAKAHASHI (1956) Hormonal control of melanin pigmentation, *Rec. Prog. Hormone Res.* **12**, 303.

LEWIS, M. R., and E. O. BUTCHER (1936) The melanophore hormone of the hypophysis cerebri of certain selachians, *Bull. Mt. Desert Island Biol. Lab.* **1936**, 20.

LI, C. H., J. S. DIXON, and D. CHUNG (1958) The structure of bovine corticotropin, *J. Amer. Chem. Soc.* **80**, 2587.

LI, C. H., I. I. GESCHWIND, R. D. COLE, I. D. RAACKI, J. I. HARRIS, and J. S. DIXON (1955) Amino-acid sequence of alpha-corticotropin, *Nature, Lond.* **176**, 687.

LUNDSTROM, H. M., and P. BARD (1932) Hypophysial control of cutaneous pigmentation in an elasmobranch fish. *Biol. Bull., Woods Hole.* **62**, 1.

MARSLAND, D. A. (1944) Mechanism of pigment displacement in unicellular chromatophores, *Biol. Bull., Woods Hole.* **87**, 252.

MARTINI, E., and I. ACHUNDOW (1929) Versuche über Farbenanpassung bei Culiciden, *Zool. Anz.* **81**, 25.

MATSUMOTO, K. (1954a) Neurosecretion in the thoracic ganglion of the crab, *Eriocheir japonicus, Biol. Bull., Woods Hole.* **106**, 60.

MATSUMOTO, K. (1954b) Chromatophorotropic activity of the neurosecretory cells in the thoracic ganglion of *Eriocheir japonicus, Biol. J. Okayama Univ.* **1**, 234.

MATSUMOTO, K. (1956) Migration of the neurosecretory products in the thoracic ganglion of the crab, *Chionoecetes opilio*, *Biol. J. Okayama Univ.* **2**, 137.

MATSUMOTO, K. (1958) Morphological studies on the neurosecretion in crabs, *Biol. J. Okayama Univ.* **4**, 103.

MATSUMOTO, K. (1959) Neurosecretory cells of an isopod, *Armadillidium vulgare* (Latreille), *Biol. J. Okayama Univ.* **5**, 43.

MATTHEWS, S. A. (1931) Observations on pigment migration within the fish melanophore, *J. Exp. Zool.* **58**, 471.

MATTHEWS, S. A. (1933) Color changes in *Fundulus* after hypophysectomy, *Biol. Bull., Woods Hole.* **64**, 315.

MAYNARD, D. M. (1961) Thoracic neurosecretory structures in Brachyura. 1. Gross anatomy. *Biol. Bull., Woods Hole* **121**, 316.

McVAY, J. A. (1942) Physiological experiments upon neurosecretion with special reference to *Lumbricus* and *Cambarus*, Doctorate Thesis, Northwestern Univ., Evanston, Illinois.

McWHINNIE, M. A., and H. M. SWEENEY (1955) The demonstration of two chromatophorotropically active substances in the land isopod, *Trachelipus rathkei*, *Biol. Bull., Woods Hole.* **108**, 160.

MEYER, G. F., and O. PFLUGFELDER (1958) Elektronenmikroskop-Untersuchungen an den Corpora Cardiaca von *Carausius morosus* Br., *Z. Zellforsch.* **48**, 556.

MILLOTT, N. (1950) The sensitivity to light, reactions to shading, pigmentation and color change of the sea urchin, *Diadema antillarum* Philippi, *Biol. Bull., Woods Hole.* **99**, 329.

MILLOTT, N. (1952) Colour change in the echinoid, *Diadema antillarum* Philippi, *Nature. Lond.* **170**, 325.

MILLOTT, N., and M. YOSHIDA (1956) Reactions to shading in the sea urchin *Psammechinus miliaris* (Gmelin), *Nature, Lond.* **178**, 1300.

MILLOTT, N., and M. YOSHIDA (1957) The spectral sensitivity of the echinoid *Diadema antillarum* Phillipi, *J. Exp. Biol.* **34**, 394.

MILLS, S. M. (1932a) Double innervation of melanophores, *Proc. Nat. Acad. Sci., Wash.* **18**, 538.

MILLS, S. M. (1932b) Neuro-humoral control of fish melanophores, *Proc. Nat. Acad. Sci., Wash.* **18**, 540.

MILLS, S. M. (1932c) The double innervation of fish melanophores, *J. Exp. Zool.* **64**, 231.

MILLS, S. M. (1932d) Evidence for a neurohumoral control of fish melanophores, *J. Exp. Zool.* **64**, 245.

MIYAWAKI, M. (1956) PAS-positive material in the neurosecretory cells of the crab, *Telmessus cheiragonus* (Tilesius), *Annot. Zool. Jap.* **29**, 151.

MIYAWAKI, M. (1960a) On the neurosecretory cells of some decapod Crustacea, *Kumamoto J. Sci., Ser. B*, **5**, 1.

MIYAWAKI, M. (1960b) Studies on the cytoplasmic globules in the nerve cells of the crab, *Gaetice depressus*. I. Histochemical observations, *Kumamoto J. Sci., Ser. B*, **5**, 21.

MIYAWAKI, M. (1960c) Studies on the cytoplasmic globules in the nerve cells of the crabs *Gaetice depressus* and *Potamon dehaani*. II. Observations by conventional, phase-contrast and electron microscopes, *Kumamoto J. Sci., Ser. B*, **5**, 29.

MORRIS, C. J. O. R. (1952) ACTH and the pigment hormone, *Lancet*, **1952** (1) 1210.

MOTHES, G. (1960) Weitere Untersuchungen über den physiologischen Farbwechsel von *Carausius morosus* (Br.), *Zool. Jber.* **69**, 133.

MÜLLER, J. (1953) Über die Wirkung von Thyroxin und thyreotropem Hormon auf den Stoffwechsel und die Farbung des Goldfisches, *Z. vergl. Physiol.* **35**, 1.

MÜSSBICHLER, A., and K. UMRATH (1950) Über den Farbwechsel von *Hyla, arborea, Z. vergl. Physiol.* **32**, 311.

NAGANO, T. (1943) Physiological studies on the pigmentary system of Crustacea. I. The color change of a shrimp *Paratya compressa* (de Haan), *Sci. Repts. Tôhoku Imp. Univ.*, 4th Ser. (*Biology*), **17**, 223.

NAGANO, T. (1949) Physiological studies on the pigmentary system of Crustacea. III. The color change of an isopod *Ligia exotica* (Roux), *Sci. Repts. Tôhoku Univ.*, 4th Ser. (*Biology*) **18**, 167.

NAGANO, T. (1950) Physiological studies on the pigmentary system of Crustacea. V. Drug action upon the pigmentary system of a shrimp, *Sci. Repts. Tôhoku Univ.*, 4th Ser. (*Biology*) **18**, 298.

NIEUWKOOP, P. D., and J. FABER (1956) Normal table of *Xenopus laevis* (Daudin), North Holland Publishing Co., Amsterdam.

NOVALES, R. R. (1959) The effects of osmotic pressure and sodium concentration on the response of melanophores to intermedin, *Physiol. Zool.* **32**, 15.

NOVALES, R. R. (1960) Responses of tissue-cultured embryonic newt melanophores to epinephrine and intermedin, *Trans. Amer. Micr. Soc.* **79**, 25.

NOVALES, R. R., and B. J. NOVALES (1961) Sodium dependence of intermedin action on melanophores in tissue culture, *Gen. Comp. Endocrinol.* **1**, 134.

NOVALES, R. R., B. J. NOVALES, and S. H. ZINNER (1960) Further studies on ionic factors influencing intermedin action on frog skin, *Anat. Rec.* **138**, 374.

OGURO, C. (1959a) Occurrence of accessory sinus gland in the isopod, *Idotea japonica, Annot. Zool. Jap.* **32**. 71.

OGURO, C. (1959b) On the sinus glands in four species belonging to the Idoteidae (Crustacea, Isopoda), *Jour. Fac. Sci. Hokkaido Univ., Ser. VI, Zool.* **14**, 261.

OGURO, C. (1959c) On the physiology of melanophores in the marine isopod, *Idotea japonica* I, *Endocrinologia Jap.* **6**, 246.

OKAY, S. (1943) Changement de coloration chez *Sphaeroma serratum* Fabr., *Rev. Fac. Sci. Univ. Istanbul, Ser. B.* **9**, (3) 204.

OKAY, S. (1945a) Sur l'excitabilité directe des chromatophores, les changements périodiques de coloration et le centre chromatophorotropique chez *Sphaeroma serratum* Fabr., *Rev. Fac. Sci. Univ. Istanbul, Ser. B,* **9**, (5), 1.

OKAY, S. (1945b) L'hormone de contraction des cellules pigmentaires chez les isopodes, *Rev. Fac. Sci. Univ. Istanbul, Ser. B,* **10**, 116.

ORTMAN, R. (1954) Cytological study of the physiological activities in the pars intermedia of *Rana pipiens, Anat. Rec.* **119**, 1.

ORTMAN, R. (1956) A study of the effect of several experimental conditions on the intermedin content and cytochemical reactions of the intermediate lobe of the frog (*Rana pipiens*), *Acta Endocrinol.* **23**, 437.

ÖSTLUND, N., and R. FÄNGE (1956) On the nature of the eye-stalk hormone which causes concentration of red pigment in shrimps (Natantia), *Ann. Sci. Nat.* **18**, 325.

OZTAN, N., and A. GORBMAN (1960) The hypophysis and hypothalamo-hypophyseal neurosecretory system of larval lampreys and their responses to light, *J. Morph.* **106**, 243.

PALAY, S. L. (1955) An electron microscope study of the neurohypophysis in normal, hydrated, and dehydrated rats, *Anat. Rec.* **121**, 348.

PANOUSE, J. B. (1946) Recherches sur les phénomènes humoraux chez les crustacés, l'adaptation chromatique et la croissance ovarienne chez la crevette *Leander serratus, Ann. Inst. Océan. Monaco.* **23**, 65.

PARKER, G. H. (1930) The color changes of the tree toad in relation to nervous and humoral control, *Proc. Nat. Acad. Sci., Wash.* **16**, 395.

PARKER, G. H. (1931) The color changes in the sea-urchin *Arbacia, Proc. Nat. Acad. Sci., Wash.* **17**, 594.

PARKER, G. H. (1933) The cellular transmission of neurohumoral substances in melanophore reactions, *Proc. Nat. Acad. Sci., Wash.* **19**, 175.

PARKER, G. H. (1934a) Cellular transfer of substances, especially neurohumors, *J. Exp. Biol.* **11**, 81.

PARKER, G. H. (1934b) Color changes in the catfish *Ameiurus* in relation to neurohumors, *J. Exp. Zool.* **69**, 199.

PARKER, G. H. (1935a) The electric stimulation of the chromatophoral nerve-fibers in the dogfish, *Biol. Bull., Woods Hole.* **68**, 1.

PARKER, G. H. (1935b) The chromatophoral neurohumors of the dogfish, *J. Gen. Physiol.* **18**, 837.

PARKER, G. H. (1935c) An oil-soluble neurohumor in the catfish *Ameiurus, J. Exp. Zool.* **12**, 239.

PARKER, G. H. (1936a) Color changes in elasmobranchs, *Proc. Nat. Acad. Sci., Wash.* **22**, 55.

PARKER, G. H. (1936b) Integumentary color changes in the newly-born dogfish, *Mustelus canis, Biol. Bull., Woods Hole.* **70**, 1.

PARKER, G. H. (1936c) The reactivation by cutting of several melanophore nerves in the dogfish *Mustelus, Biol. Bull., Woods Hole.* **71**, 255.

PARKER, G. H. (1937) Integumentary color changes of elasmobranch fishes, especially of *Mustelus, Proc. Amer. Phil. Soc.* **77**, 223.

PARKER, G. H. (1938a) Melanophore responses and blood supply (vasomotor changes), *Proc. Amer. Phil. Soc.* **78**, 513.

PARKER, G. H. (1938b) The colour changes in lizards, particularly in *Phrynosoma, J. Exp. Biol.* **15**, 48.

PARKER, G. H. (1940) The chromatophore system in the catfish *Ameiurus, Biol. Bull., Woods Hole.* **79**, 237.

PARKER, G. H. (1941) Melanophore bands and areas due to nerve cutting in relation to the protracted activity of nerves, *J. Gen. Physiol.* **24**, 483.

PARKER, G. H. (1942) Color changes in *Mustelus* and other elasmobranch fishes, *J. Exp. Zool.* **89**, 451.

PARKER, G. H. (1943) Animal color changes and their neurohumors, *Quart. Rev. Biol.* **18**, 205.

PARKER, G. H. (1945) Melanophore activators in the common American eel *Anguilla rostrata* Le Sueur, *J. Exp. Zool.* **98**, 211.

PARKER, G. H. (1948) *Animal Colour Changes and Their Neurohumours*, Cambridge Univ. Press, Cambridge.

PARKER, G. H., and H. P. BROWER (1937) An attempt to fatigue the melanophore system in *Fundulus* and a consideration of lag in melanophore responses, *J. Cell. Comp. Physiol.* **9**, 315.

PARKER, G. H., and H. PORTER (1934) The control of the dermal melanophores in elasmobranch fishes, *Biol. Bull., Woods Hole.* **66**, 30.

PARKER, G. H., and A. ROSENBLUETH (1941) The electric stimulation of the concentrating (adrenergic) and the dispersing (cholinergic) nerve-fibers of the melanophores in the catfish, *Proc. Nat. Acad. Sci., Wash.* **27**, 198.

PARKER, G. H., and L. E. SCATTERTY (1937) The number of neurohumors in the control of frog melanophores, *J. Cell. Comp. Physiol.* **9**, 297.

PASSANO, L. M. (1951a) The X-organ sinus gland neurosecretory system in crabs, *Anat. Rec.* **111**, 502.

PASSANO, L. M. (1951b) The X-organ, a neurosecretory gland controlling molting in crabs, *Anat. Rec.* **111**, 559.

PASSANO, L. M. (1953) Neurosecretory control of molting in crabs by the X-organ sinus gland complex, *Physiol. Comp. Oecol.* **3**, 155.

PASTEUR, C. (1958) Influence de l'ablation de l'organe X sur le comportment chromatique de *Leander serratus* (Pennant), *C. R. Acad. Sci., Paris.* **246**, 320.

PATTON, J. R., and R. S. TEAGUE (1959) The response of *Hyla cinerea* to pituitary intermedin as measured by a reflection technique, *Honoris Causae Volume for Prof. G. Joachimoglu*, Monotypias School Press, Athens.

PAUTSCH, F. (1952) Chromatophorotropic principles of the walking stick *Dixippus morosus* Br. and Redt. as colour change activators in some amphibians and marine crustaceans, *Bull. Acad. Polon. Sci. Lettres. Ser. B.* **1951**, 17.

PAUTSCH, F. (1953) Colour adaptations of the zoea of the shrimp *Crangon crangon* L., *Bull. Acad. Polon. Sci. Lettres, Ser. B.* **1951**, 511.

PÉREZ-GONZÁLEZ, M. D. (1957) Evidence for hormone-containing granules in sinus glands of the fiddler crab *Uca pugilator*, *Biol. Bull., Woods Hole.* **113**, 426.

PERKINS, E. B. (1928) Color changes in crustaceans, especially in *Palaemonetes*, *J. Exp. Zool.* **50**, 71.

PERKINS, E. B., and B. KROPP (1932) The crustacean eye hormone as a vertebrate melanophore activator, *Biol. Bull., Woods Hole.* **63**, 108.

PICKFORD, G. E. (1956) Melanogenesis in *Fundulus heteroclitus*, *Anat. Rec.* **125**, 603.

PICKFORD, G. E., and J. W. ATZ (1957) *The Physiology of the Pituitary Gland of Fishes*, New York Zoological Society, New York.

POTTER, D. D. (1954) Histology of the neurosecretory system of the blue crab *Callinectes sapidus*, *Anat. Rec.* **120**, 716.

POTTER, D. D. (1958) Observations on the neurosecretory system of portunid crabs, *Zweites Internat. Sympos. über Neurosekretion*, W. BARGMANN, B. HANSTRÖM, B. SCHARRER, and E. SCHARRER, eds., Springer-Verlag, Berlin.

PROSSER, C. L., and F. A. BROWN, JR. (1961) *Comparative Animal Physiology*, W. B. Saunders Co., Philadelphia.

RAHN, H. (1941) The pituitary regulation of melanophores in the rattlesnake, *Biol. Bull., Woods Hole.* **80**, 228.

RAHN, H. (1956) The relationship between hypoxia, temperature, adrenalin release and melanophore expansion in the lizard, *Anolis carolinensis*, *Copeia.* **1956**, 214.

RASQUIN, P. (1958) Studies in the control of pigment cells and light reaction in recent teleost fishes. II. Reactions of the pigmentary system to hormonal stimulation, *Bull. Amer. Mus. Nat. Hist.* **115**, 34.

REDFIELD, A. C. (1918) The physiology of the melanophores of the horned toad *Phrynosoma*, *J. Exp. Zool.* **26**, 275.

REIDINGER, L. (1952) Über den morphologischen und physiologischen Farbwechsel der Elritze, *Z. vergl. Physiol.* **34**, 394.

REINDEL, F., and W. HOPPE (1954) Über eine Färbemethode zum Anfarben von Aminosäuren, Peptiden und Proteinen auf Papierchromatogrammen und Papierelektrogrammen, *Chem. Ber.* **87**, 1103.

RIGLER, R., and M. HOLZBAUER (1953) Untersuchungen über die wechselseitigen Beziehungen von ACTH zu Intermedinwirkung an der Froschhaut, *Arch. Exp. Path. Pharmak.* **219**, 456.

ROBERTSON, O. H. (1951a) Factors influencing the state of dispersion of the dermal melanophores in rainbow trout, *Physiol. Zool.* **24**, 309.

ROBERTSON, O. H. (1951b) The relationship of muscle potassium to the melanophore-concentrating effect of pressure on the trout, *Science*, **114**, 11.

ROUNDS, H. D., and M. McCLAIN (1961) Segmental distribution of insect chromatophorotropins, *Amer. Zoologist*, **1**, 383.

ROWLANDS, A. (1950) The influence of water and light upon the pigmentary system in the common frog, *Rana temporaria*, *J. Exp. Biol.* **27**, 446.

ROWLANDS, A. (1952) The influence of water and light upon the colour change of sightless frogs (*Rana temporaria*), *J. Exp. Biol.* **29**, 127.

ROWLANDS, A. (1954) The influence of water and light and the pituitary upon the pigmentary system of the common toad (*Bufo bufo bufo*), *J. Exp. Biol.* **31**, 151.

SANDEEN, M. I. (1950) Chromatophorotropins in the central nervous system of *Uca pugilator*, with special reference to their origins and actions, *Physiol. Zool.* **23**, 337.

SANDEEN, M. I., and J. D. COSTLOW, JR. (1961) The presence of decapod-pigment-activating substances in the central nervous system of representative Cirripedia, *Biol. Bull., Woods Hole.* **120**, 192.

SANDEEN, M. I., and M. FINGERMAN (1959) Studies by filter paper electrophoresis of the hormones controlling color changes in the shrimp *Crangon*, *Biol. Bull., Woods Hole.* **117**, 425.

SCHAEFER, J. G. (1921) Beitrage zur Physiologie des Farbwechsels der Fische. I. Untersuchungen an Pleuronectiden. II. Weitere Untersuchungen, *Pflüg. Arch. Ges. Physiol.* **188**, 25.

SCHARRER, B. (1941) Neurosecretion. IV. Localization of neurosecretory cells in the central nervous system of *Limulus*, *Biol. Bull., Woods Hole.* **81**, 96.

SCHARRER, B. (1952a) Hormones in Insects, *The Action of Hormones in Plants and Invertebrates*, K. V. THIMANN, ed., Academic Press Inc., New York.

SCHARRER, B. (1952b) Neurosecretion. XI. The effects of nerve section on the intercerebralis-cardiacum-allatum system of the insect *Leucophaea maderae*, *Biol. Bull., Woods Hole.* **102**, 261.

SCHARRER, E., and B. SCHARRER (1954) *Neurosecretion, Handbuch der Mikroskopischen Anatomie des Menschen*, U. MOLLENDORK and W. BARGMANN, eds., SPRINGER-VERLAG, Berlin, **6**, 953.

SCHEER, B. T. (1960) The neuroendocrine system of arthropods, *Vitamins and Hormones*, **18**, 141.

SCHEER, B. T., and M. A. R. SCHEER (1954) The hormonal control of metabolism in crustaceans. VII. Moulting and colour change in the prawn *Leander serratus*, *Publ. Staz. Zool. Napoli.* **25**, 397.

SCHEER, B. T., and M. A. R. SCHEER (1955) Relation of colour change to moulting in prawns, *Nature, Lond.* **175**, 473.

SCHWINCK, I. (1953) Über den Nachweis eines Redox-Pigmentes (Ommochrom) in der Haut von *Sepia officinalis*, *Naturwissenschaften.* **40**, 365.

SCHWINCK, I. (1955) Vergleich des Redox-Pigmentes aus Chromatophoren und Retina von *Sepia officinalis* mit Insektenpigmenten der Ommochromgruppe, *Verh. Deutsch. Zool. Ges. Erlangen, 1955, Zool. Anz. Suppl.* **19**, 71.

SCHWYZER, R., and C. H. LI (1958) A new synthesis of the pentapeptide l-histidyl-l-phenylalanyl-l-arginyl-l-tryptophyl-glycine and its melanocytestimulating activity, *Nature, Lond.* **182**, 1669.

SERENI, E. (1928) Sui cromatofori dei cefalopodi. I. Azione di alcuni veleni *in vivo, Z. vergl. Physiol.* **8**, 488.

SERENI, E. (1929a) Sulla funzione delle ghiandole salivari posteriori dei cefalopodi, *Boll. Soc. Ital. Biol. Sper.* **4**, 736.

SERENI, E. (1929b) Correlazione umorali nei cefalopodi, *Amer. J. Physiol.* **90**, 512.

SERENI, E. (1929c) Sul meccanismo d'azione della veratrina, *Boll. Soc. Ital. Biol. Sper.* **4**, 1211.

SERENI, E. (1930a) Sui cromatofori dei cefalopodi. III. Azione di alcuni veleni *in vitro, Z. vergl. Physiol.* **12**, 329.

SERENI, E. (1930b) The chromatophores of the cephalopods, *Biol. Bull., Woods Hole.* **59**, 247.

SERENI, E. (1930c) Sui cromatofori dei cefalopodi. II. Azione della betaine e della arecolina, *Arch. Farmacol. Sper.* **48**, 223.

SHEPHERD, R. G., S. D. WILLSON, K. S. HOWARD, D. H. BELL, D. S. DAVIES, E. H. EIGNER, and N. E. SHAKESPEARE (1956) Studies with corticotropin. III. Determination of the structure of B-corticotropin and its active degradation products, *J. Amer. Chem. Soc.* **78**, 5067.

SHIZUME, K., A. B. LERNER, and T. B. FITZPATRICK (1954) *In vitro* bioassay for the melanocyte stimulating hormone, *Endocrinology* **54**, 553.

SIEDLECKI, M. (1909) Zur Kenntnis des javanischen Flugfrosches, *Biol. Centr.* **29**, 704.

SIEGLITZ, G. (1951) Über hormonale und nervöse Regulation des Farbwechsels beim Frosch (*Rana temporaria*), *Z. vergl. Physiol.* **33**, 99.

SMITH, H. G. (1938) Receptive mechanism of background response in chromatic behaviour of crustacea, *Proc. Roy. Soc., B.* **125**, 250.

SPAETH, R. A. (1913) The mechanism of the contraction in the melanophores of fishes, *Anat. Anz.* **44**, 520.

STÅHL, F. (1938a) Preliminary report on the colour changes and the incretory organs in the heads of some crustaceans, *Ark. Zool.* **30B**, 1.

STÅHL, F. (1938b) Über das Vorkommen von inkretorischen Organen und Farbwechsel-hormonen im Kopf einiger Crustaceen, *Lunds Univ. Årsskrift,* **34**, 1.

168 THE CONTROL OF CHROMATOPHORES

STEGGERDA, F. R., and A. L. SODERWALL (1939) Relationship of the pars tuberalis to melanophore response in Amphibia (*Rana pipiens*), *J. Cell. Comp. Physiol.* **13**, 31.

STEPHENS, G. C. (1957a) Twenty-four hour cycles in marine organisms, *Amer. Nat.* **111**, 135.

STEPHENS, G. C. (1957b) Influence of temperature fluctuations on the diurnal melanophore rhythm of the fiddler crab *Uca*, *Physiol. Zool.* **30**, 55.

STEPHENS, G. C., M. F. FRIEDL, and B. GUTTMAN (1956) Electrophoretic separation of chromatophorotropic principles of the fiddler crab, *Uca*, *Biol. Bull. Woods Hole.* **111**, 312.

STOPPANI, A. O. M. (1941) La regulación nerviosa del color del *Bufo arenarum* Hensel, *Rev. Soc. Argent. Biol.* **17**, 484.

STOPPANI, A. O. M. (1942) Neuroendocrine mechanism of color change in *Bufo arenarum* Hensel, *Endocrinology.* **30**, 782.

STOPPANI, A. O. M., P. F. PIERONI, and A. J. MURRAY (1954) The role of peripheral nervous system in colour changes of *Bujo arenarum* Hensel, *J. Exp. Biol.* **31**, 631.

STROHM, F. (1936) Über chemische Beeinflussung der Froschhautmelanophoren, *Med. Diss. Tübingen 1936.*

SULMAN, F. G. (1952a) Chromatophorotropic effect of adrenocorticotropic hormone, *Nature, Lond.* **169**, 588.

SULMAN, F. G. (1952b) The effect of ACTH on the frog chromatophore, *Acta Endocrinol.* **10**, 320.

SUMNER, F. B. (1933) The differing effects of different parts of the visual field upon the chromatophore responses of fishes, *Biol. Bull. Woods Hole.* **65**, 266.

SUMNER, F. B. (1935) Evidence for the protection value of changeable coloration in fishes, *Amer. Nat.* **69**, 245.

SUNESON, S. (1947) Colour change and chromatophore activators in *Idothea*, *Lunds Univ. Årsskrift*, **43**, 5.

TAYLOR, A., and J. J. KOLLROS (1946) Stages in the normal development of *Rana pipiens* larvae, *Anat. Rec.* **94**, 7.

TEAGUE, R. S., and J. R. PATTON (1960) Analysis of the spectrophotometric reflectance response of frogs to melanophore hormone, *J. Cell. Comp. Physiol.* **56**, 15.

TEISSIER, G. (1947) Fonctionnement des chromatophores de la larve de Corèthre, *C. R. Acad. Sci., Paris.* **225**, 204.

THING, E. (1952) Melanophore reaction and adrenocorticotrophic hormone. I. Comparison of methods based upon microscopic observations of the melanophores, *Acta Endocrinol.* **10**, 295.

THOMSEN, E. (1954) Experimental evidence for the transport of secretory material in the axons of the neurosecretory cells of *Calliphora erythrocephala* Meig, *Pubbl. Staz. Zocl. Napoli Suppl.* **24**, 48.

TRAVIS, D. F. (1951) The control of the sinus glands over certain aspects of calcium metabolism in *Panulirus argus*, *Anat. Rec.* **111**, 503.

UMRATH, K. (1957) Über den physiologischen und den morphologischen Farbwechsel des Bitterlings, *Rhodeus amarus*, *Z. vergl. Physiol.* **40**, 321.

UMRATH, K. (1959) Über den Einfluss des adrenocorticotropen Hormons auf die Färbung und über die Auslösbarkeit des Hochzeitkleides bei einigen Fischen, *Z. vergl. Physiol.* **42**, 181.

UMRATH, K., and H. WALCHER (1951) Farbwechselversuche an *Macropodus opercularis* und ein Vergleich der Geschwindigkeit der Farbänderung bei Macropoden und Elritzen, *Z. vergl. Physiol.* **33**, 129.

UNGER H. (1960) Neurohormone bei Seesternen (*Marthasterias glacialis*), *Symp. Biol. Hung.* **1**, 203.

VAN DER LEK, B., J. DE HEER, A. C. J. BURGERS, and C. J. VAN OORDT (1958) The direct reaction of the tail fin melanophores to light, *Acta Physiol. Pharmacol. Neerlandica*, **7**, 409.

VAN DE VEERDONK, F. C. G (.1960) Serotonin, a melanocyte-stimulating component in the dorsal skin secretion of *Xenopus laevis*, *Nature, Lond.* **187**, 948.

VAN DE VEERDONK, F. C. G., J. W. HUISMANS, and A. D. F. ADDINK (1961) A melanocyte-stimulating substance in the skin secretion of *Xenopus laevis*, *Z. vergl. Physiol.* **44**, 323.

VAN RYNBERK, G. (1906) Über den durch Chromatophoren bedingten Farbenwechsel der Tiere (sog. chromatische Hautfunktion), *Ergebn. Physiol.* **5**, 347.

VEIL, C., and R. M. MAY (1937) Hypophysectomie et changement de couleur chez la torpille (*Torpedo marmorata*), *C. R. Soc. Biol., Paris.* **124**, 917.

VILTER, V. (1937) Recherches histologiques et physiologiques sur la fonction pigmentaire des sélaciens, *Bull. Soc. Sci. Arcachon* **34**, 65.

VOLPE, E. P., and J. L. DOBIE (1959) The larva of the oak toad, *Bufo quercicus*, *Tulane Studies Zool.* **7**, 145.

VOLPE, E. P., and S. M. HARVEY (1958) Hybridization and larval development in *Rana palmipes* Spix, *Copeia*, **1958**, 197.

VOLPE, E. P., M. A. WILKENS, and J. L. DOBIE (1961) Embryonic and larval development of *Hyla avivoca*, *Copeia*, **1961**, 340.

VON FRISCH, K. (1911) Beiträge zur Physiologie der Pigmentzellen in der Fischhaut, *Pflüg. Arch. Ges. Physiol.* **138**, 319.

VON GELEI, G. (1942) Zur Frage der Doppelinnervation der Chromatophoren, *Z. vergl. Physiol.* **29**, 532.

VON UEXKÜLL, J. (1896) Vergleichend sinnesphysiologische Untersuchungen. II. Der Schatten als Reiz für *Centrostephanus longispinus*, *Z. Biol.* **34**, 319.

WARING, H. (1936) Color changes in the dogfish (*Scyllium canicula*), *Trans. Lpool. Biol. Soc.* **49**, 17.

WARING, H. (1938) Chromatic behaviour of elasmobranchs, *Proc. Roy. Soc. B.* **125**, 264.

WARING, H. (1940) The chromatic behaviour of the eel (*Anguilla vulgaris* L.), *Proc. Roy. Soc. B.* **128**, 343.

WARING, H., and F. W. LANDGREBE (1950) Hormones of the posterior pituitary, *The Hormones Vol. II*, G. PINCUS and K. V. THIMANN, eds., Academic Press, Inc., New York.

WEBB, H. M. (1950) Diurnal variations of response to light in the fiddler crab, *Uca, Physiol. Zool.* **23**, 316.

WEBB, H. M., M. F. BENNETT, and F. A. BROWN, JR. (1954) A persistent diurnal rhythm of chromatophoric response in eyestalkless *Uca pugilator*, *Biol. Bull. Woods Hole.* **106**, 371.

WEBB, H. M., and F. A. BROWN, JR. (1959) Timing long-cycle physiological rhythms, *Physiol. Rev.* **39**, 127.

WEBB, H. M., F. A. BROWN, JR., M. F. BENNETT, J. SHRINER, and R. A. BROWN (1956) An alteration of the persistent daily rhythm of color change of fiddler crabs effected by isolation, *Anat. Rec.* **125**, 615.

WEBB, H. M., F. A. BROWN, Jr., and R. C. GRAVES (1952) The influence of light intensity and temperature on chromatophores of *Palaemonetes vulgaris, Biol. Bull. Woods Hole.* **103**, 310.

WEISEL, G. F. (1950) The comparative effects of teleost and beef pituitaries on chromatophores of cold-blooded vertebrates, *Biol. Bull. Woods Hole.* **99**, 487.

WELSH, J. H. (1941) The sinus gland and 24-hour cycles of retinal pigment migration in the crayfish, *Cambarus bartoni, J. Exp. Zool.* **86**, 35.

WELSH, J. H. (1951) New evidence concerning the source and action of the eyestalk hormone, *Anat. Rec.* **111**, 442.

WHITE, W. F., and W. A. LANDMANN (1955) Studies of adrenocorticotropin, XI. A preliminary comparison of corticotropin-A with B-corticotropin, *J. Amer. Chem. Soc.* **77**, 1711.

WOOLLEY, P. (1957) Colour change in a chelonian, *Nature, Lond.* **179**, 1255.

WRIGHT, M. R., and A. B. LERNER (1960a) On the movement of pigment granules in frog melanocytes, *Endocrinology.* **66**, 599.

WRIGHT, M. R., and A. B. LERNER (1960b) Action of thyroxine analogues on frog melanocytes, *Nature, Lond.* **185**, 169.

WRIGHT, P. A. (1948) Photoelectric measurement of melanophoral activity of frog skin induced *in vitro, J. Cell. Comp. Physiol.* **31**, 111.

WRIGHT, P. A. (1954a) A convenient and rapid technique for assay of intermedin, *Papers Mich. Acad. Sci., Arts and Letters* **39**, 271.

WRIGHT, P. A. (1954b) Inhibition of intermedin-induced darkening of frog skin (*Rana pipiens*) by tetrazolium salt, *Biol. Bull. Woods Hole* **107**, 323.

WRIGHT, P. A. (1955) Physiological responses of frog melanophores *in vitro, Physiol. Zool.* **28**, 204.

WYKES, V. (1936) Observations on pigmentary coordination in elasmobranchs, *J. Exp. Biol.* **13**, 460.

YOSHIDA, M. (1956) On the light responses of the chromatophores of the sea-urchin, *Diadema setosum* (Leske), *J. Exp. Biol.* **33**, 119.

YOSHIDA, M. (1957) Spectral sensitivity of chromatophores in *Diadema setosum* (Leske), *J. Exp. Biol.* **34**, 222.

YOUNG, J. Z. (1933) The autonomic nervous system of selachians, *Quart. J. Micr. Sci.* **75**, 571.

YOUNG, J. Z. (1935) The photoreceptors of lampreys, II. The functions of the pineal complex, *J. Exp. Biol.* **12**, 254.

YOUNG, J. Z., and C. W. BELLERBY (1935) The response of the lamprey to injection of anterior lobe pituitary extract, *J. Exp. Biol.* **12**, 246.

ZOOND, A., and J. EYRE (1934) Studies in reptilian colour response, I. The bionomics and physiology of the pigmentary activity of the chameleon, *Phil. Trans., B.* **223**, 27.

AUTHOR INDEX

173

SUBJECT INDEX

178